THE IRAQ LEVIES

1915-1932

Early 1916. 1916–1917. Late 1917.

THE IRAQ LEVIES.

THE IRAQ LEVIES

1915-1932

By

BRIGADIER J. GILBERT BROWNE

C.M.G., C.B.E., D.S.O.

1932

LONDON

THE ROYAL UNITED SERVICE INSTITUTION

WHITEHALL, S.W.

PREFACE

THIS is a short account of the work done by the Iraq Levies, from their beginning in 1915, to their final change over into the New Air Defence Force, in 1932.

I am indebted to the following for letting me have information, which has helped to compile this account :—

Lieut.-Colonel G. C. M. Sorel Cameron, C.B.E.
Lieut.-Colonel C. R. Barke, C.B.E.
Mr. Edmonds, of the Political Branch, High Commissioners' Staff, Iraq.
Lieut.-Colonel L. W. Alexander, C.B.E.
Major T. P. Aldworth, D.S.O., O.B.E.
Major F. W. Hall, M.B.E.
Major C. A. Boyle, D.S.O.
Captain R. Merry, M.B.E.
Captain C. E. Littledale, O.B.E., M.C., Iraq Police.
Captain P. J. T. Baddiley, M.B.E.
Captain C. E. P. Hooker.
Captain F. J. McWhinnie.
Lieutenant C. O. L. Devenish.
R.S.M. E. G. Walker, M.B.E.
C.Q.M.S. C. Chapman.

Finally I am also greatly indebted to Mr. K. R. Wilson for the trouble he has taken in reading the proofs and for his assistance in the practical production of this History.

J. GILBERT BROWNE,
Brigadier.

HINAIDI,
12th July, 1932.

CONTENTS

APPENDICES

ILLUSTRATIONS

MAPS

1927.

THE IRAQ LEVIES.

THE IRAQ LEVIES

CHAPTER I

1915–1919

IN 1915, Major J. I. Eadie * of the Indian Army, who was then Special Service Officer in the Muntafiq Division in Mesopotamia, recruited forty Mounted Arabs from the tribes round Nasiriyeh, on the Euphrates, for duty under the Intelligence Department. 1915.

From this small force of forty men was gradually built up a force, which, after various changes of name, were finally called "Levies"; and which, from a strength of 40 in 1915, rose to 6,199 in May 1922, after which date the gradual cutting down of units, or transfer to the Iraq Army, began.

The following is an attempt to give an account of this force, whose organization changed from a small mounted contingent to a mixed force of all arms; whose personnel changed from entirely Arabs, to a mixed force of Arabs, Kurds, Assyrians, Turkomans and Yezidis, and finally to almost entirely Assyrians; and whose area of use was first of all limited entirely to the country south of Baghdad, later entirely to Kurdistan, and now, as their end approaches, they are gradually taking over stations in the South again.

Major Eadie's forty men, at first known locally as the Muntafiq Horse, were soon increased to sixty, and were called Arab Scouts. Their duties were many and various, and included reconnoitring for British columns which were operating in the area. They were allowed to wear their own form of dress, produced their own horses, saddlery, rifles, arms and ammunition, and provided their own shelter for themselves and their animals. They formed the nucleus of the 5th Euphrates Levy. Arab Scouts.

In March 1916, another small mounted force, also sixty strong, was raised by Major Hamilton, the Political Officer at Nasiriyeh. This was called the Political Guard. This force acted as guard to the Political 1916, March. Political Guard.

* Later Lieut.-Colonel J. I. Eadie, D.S.O.

Officer during his tours of the Division, and carried out Police duties in the town and district. They were paid at the same rates, and enlisted under the same conditions as the Arab Scouts.

June. In June 1916, after the Fall of Kut, a corps of guards for the river and telegraph line in the Qurnah, Amarah and Basrah areas, was raised. This corps was divided to correspond to Political Divisions, and acted under Assistant Political Officers. These formed the nucleus of the 3rd Tigris Levy, and the Qurnah District Police.

Nasiriyeh Mounted Guard. In this same month the Arab Scouts and the Political Guard were amalgamated, and renamed the Nasiriyeh Mounted Guard. The strength of this force was raised from 120 to 150, and by the end of the year to 250.

1917, April, July. A further increase to 350, took place in April 1917, and by July, by adding other forces raised by the Civil Commissioners, the force had become 500 mounted men, and 400 dismounted men.

In 1916 it was decided that the force must have a distinctive uniform, and here a difficulty arose. The only clothing for the legs available was either shorts or riding breeches, and the Arabs refused to wear either, considering that such an exposure of their lower limbs indecent, and against all custom. A compromise was effected by which they wore their Arab robes over the uniform. However, by 1917 they became used to wearing the dress of the British soldier, and this was adopted throughout the force.

1918, April 12th. The force continued to grow, and seems to have had different names in different areas. Thus we find on 12th April 1918, that the Hillah Shabana or 2nd Euphrates Levy under Major C. A. Boyle,* were used; this being

Madhatiyah. the first operation in which this force is mentioned. This was to Madhatiyah, to destroy towns and forts, collect revenue, and make certain arrests. They covered one hundred and ten miles on this expedition without any horse casualties, destroyed eighty-four towers, collected some of the revenue and obtained surety for the rest, and brought in eight people who were wanted.

April 21st. Jerbonieh. On the 21st April the same Levy, seventy mounted men and one hundred dismounted, made an expedition to Jerbonieh and destroyed twenty-seven towers and burned two villages. The expedition lasted three days. The mounted men were eleven hours in the saddle each day. No one in the force fell out.

The Great War came to an end; but the work of this force increased.

Shabana. Their name at the end of 1918 was changed to Shabana, a name in use in Turkish times, and already in use to a certain extent. The duties were now to supply the executive needs of the Civil Administration. The

* Later Major C. A. Boyle, D.S.O.

organization, administration, and pay of the force, and strength of 5,467 were laid down in a memorandum of 8th October 1918. It was intended that the Shabana should be a striking force, and should become the nucleus of the future Arab Army. October 8th.

The name Shabana was reminiscent of many abuses in Turkish times, and generally unpopular, and the name of the force was again changed to Muntafiq Horse. And this name was again changed, the whole force in March 1919 being named Militia, though the name Muntafiq Horse seems to have continued as the name of the Mounted troops of that area. Major C. A. Boyle * was Inspecting Officer of the Militia with Headquarters at Baghdad. 1919. Muntafiq Horse. March. Militia.

It was during this year that a standard uniform for the force was laid down, the force was re-armed with the short British ·303 rifle, and a voluntary system of recruiting was introduced. So far local Sheikhs and headmen had been called on to produce men, and those produced were not exactly voluntary soldiers.

This year was fairly eventful, and the force saw a good deal of service. Before narrating this, it is best to show the changes in organization, administration, and location of the force.

In July 1919, the name of the force was changed again from Militia to Levies, in use now for the first time, and on August 1st the Levy and Gendarmerie Orders were published. These orders defined control of the Levies, and the duties of the Inspecting Officer of the Levies, which were limited to inspection and administration. July. Levies. August 1st.

Therefore by these Orders, Levies were under the orders of three different people :—

1. The Inspecting Officer.
2. The Political Officer of the Area.
3. The local Administrative Commandant.

Moreover the budget was dealt with by the Inspecting Officer, except in the Kirkuk, Sulaimani and Mosul Liwas, where Political Officers dealt with it.

This division of control, both financially and in administrative matters, was a great disadvantage. The control of the Inspecting Officer of the force over his troops, varied very much with the personality of the Political Officer in whose area they were.

Levy Headquarters began to expand, and "A" and "Q" Branch were formed in September. The force was divided into two parts :— September.

(*a*) A Striking Force at Headquarters of the Administrative Area. This numbered 3,075 and was under Levy Headquarters for training

* Later Major C. A. Boyle, D.S.O.

and other purposes. It was split up into detachments of varying sizes, throughout the country.

(*b*) District Police numbering 1,786, under the Political Officers, and only inspected from time to time by Levy Headquarters.

August 12th.

On 12th August, 1919, the force changed its name for the eighth time, becoming Arab and Kurdish Levies.

Arab and Kurdish Levies.

A memorandum was issued giving administrative details, which were however only brought into effect a year later. The chief point was that three Deputy Inspector-Generals were appointed, and for purposes of command and administration, the country was divided into three Levy areas, each under its respective Inspecting Officer, Deputy Inspecting Officer, or an Assistant Inspecting Officer, with a Staff Captain, and an Orderly Officer to assist him. The areas were :—

"A" Area Headquarters	Hillah.
"B" Area ,, ,, ...	Baghdad.
"C" Area ,, ,, ...	Mosul.

At the time this memorandum was drawn up the Levies were run by an Inspecting Officer and a small staff consisting of an Adjutant and Quartermaster, and no actual change took place in this arrangement for the present, nor was the area system brought into effect for a year.

This memorandum also laid down the general circumstances in which the Levies could be called on. For minor operations the Political Officer could call on them; but if the duty for which he wanted them involved absence from their post for twenty-four hours, permission from the Inspecting Officer of the area was required. Also no action likely to involve the force in definite hostilities was to be undertaken without reference to the Military Area Commander.

August 12th.

The organization of the force was also laid down. Mounted Levies were in Squadrons of 100, and Troops of 25. Dismounted in Companies of 100 and Platoons of 25. The memorandum gave the names and locations of units as below; but, in certain areas, such as at Samara and Khanaqin, this does not seem to have been carried into effect.

1st Euphrates Levy	Dulaim.
2nd ,, ,, ...	Hillah.
3rd ,, ,, ...	Shamiyah.
4th ,, ,, ...	Diwaniyeh.
5th ,, ,, ...	Nasiriyeh.
1st Tigris Levy	Samara.
2nd ,, ,, ...	Kut.

3rd Tigris Levy	Amara.
Deir-ez-Zor Levy	Deir-ez-Zor.
Baqubah Levy	Baqubah.
Khanaqin ,, ...	Khanaqin.
Zobeir ,, ...	Basrah.
Kirkuk ,, ...	Kirkuk.
Sulaimani ,, ...	Sulaimani.
Mosul Gendarmerie	Mosul.

It was intended also to have a new Levy at Suq es Shuyukh to keep order in the marshes, but this never appears to have been raised.

The personnel of the force at this time was drawn from:

ARABS. Mainly townspeople or from settled tribes. The desert tribes did not take kindly to discipline. A few old Arab officers of the Turkish Army also joined up.

KURDS. These joined chiefly the Sulaimani and Arbil Levies, and the Mosul Gendarmerie.

KIRKUKLIS. These are Turkoman people and joined the Kut, Baqubah and Kirkuk Levies.

The training of the Levies was laid down in a memorandum issued in November 1919. They were ordered to be practised in rapid advances, flank attacks, advanced and rear-guard action, and marsh fighting. For work in the river areas, work with aeroplanes, armoured cars, and gun-boats. They had also to be trained for mountain warfare for the northern areas, and for desert warfare anywhere to the west of Iraq. November.

One difficulty was to find officers for this force. The Sheikhly class did not take kindly to the discipline required compared with their own free life, and those who did come were found, as a rule, unsuitable. Therefore the majority of officers of the Arab Levies were promoted from the ranks, and they were not far ahead of their own men in training and experience. This affected their prestige and powers of command.

I will now give an account of the work of the Levies during this year. The first operation was on February 25th, when the 5th Euphrates Levy under Captain F. W. Hall left their station at midnight to deal with Sheikh Badr and his following, in co-operation with aeroplanes and gun-boats. The Levies numbered one hundred and twenty. February 25th.

Operation against Sheikh Badr.

It was pitch dark, and the route lay across country. Men fell into deep ditches, and one man went, horse and all, down a well. There was a lake, two and a half miles wide, between the troops and their objective, and fifty men were sent across it in bellums (a large native rowing boat), the remainder, one man to two horses, swimming and wading further down.

Orders were to remain near and in observation of the village while it was bombed and shelled, but they were not to attack unless Sheikh Badr was seen evacuating it. Therefore they remained in observation for the whole of the 26th; but pushed in and occupied the place on the 27th.

February 26th.
February 27th.
May 8th.

On May 8th thirty Levies under Captain Lewis were sent to deal with bad characters in the marsh village of Umm-el Batouch.

Umm-el Batouch.

They were transported to the island on which the village stands, on rafts, and rushed it at dawn. Five of the enemy were killed and one captured. The Levy guide was killed.

May 21st.

Sharaish River.

On May 21st the 5th Euphrates Levy were out again, forming part of a column to deal with Sheikh Badr's force, which had concentrated on the west bank of the Sharaish River. Captain Hall was again in command of the mounted troops. They marched at 11.45 p.m. on May 21st, forded the river opposite Fathi Fort, with orders to work round the north-east side of the enemy's position, and to attack while the main position was bombed with aeroplanes.

They were not to cross the Sharaish River until after the aeroplane bombardment. For some time therefore the Levies were engaged at about 900 yards with the enemy, who had advanced to the river. As soon as the bombardment ceased, they crossed, attacked, and dispersed the enemy, and burnt Badr's village and crops. The column then withdrew, the Levies protecting its rear and right flank.

On the same date matters came to a head in Sulaimani.

May 22nd.
Sheikh Mahmud's First Rebellion.
Azmir Dagh.

After a long period during which matters were getting more and more difficult, Sheikh Mahmud, who will appear very frequently later, advanced from Barzinjah on Sulaimani. He met the Levies under Major F. S. Greenhouse on May 22nd at the Azmir Dagh, overwhelmed them and captured the town.

Tasluja Pass.
Action at Bazian Pass.
June 17th.

A force in armoured cars and Fords, attempting to relieve it, was caught, and defeated with some loss in Tasluja Pass, and a very threatening situation was only brought to an end by the defeat and capture of Sheikh Mahmud in the engagement in the Bazian Pass by Major-General Sir T. Fraser, K.C.B., C.S.I., C.M.G., G.O.C., 18th Division, on the 17th June.

June.

July 15th.
Amadiyah Revolt.

The Amadiyah area was the next area of disturbance. In June the troops about Amadiyah were withdrawn to Suwara Tuka Pass, eighteen miles south-west of Amadiyah. The A.P.O., Captain Willey, was left in Amadiyah, with Lieut. MacDonald and Sergeant Troop in command of the Kurdish Levies. Anti-British and anti-Christian propaganda had been going on for some time, and on July 15th, the leaders of this movement, aided by contingents of Kurdish tribesmen and the local gendarmerie, murdered the whole party. A column at once took action under General

Nightingale, and with it two Battalions of Assyrians trained in Baqubah. They entered Amadiyah on August 8th and then took action against the Barwari tribes, the Goyan and Guli. The Assyrian Battalions did well on this expedition, and this led later to their being taken as the main part of the Iraq Levies. Operations concluded in September. **August 8th.** **September.**

Unrest in the North now extended to the Rowanduz area, where a small party of Police, with the Civil Officials, was surrounded. **Rowanduz.**

Captain C. E. Littledale left Arbil to attempt the relief of the place, having with him only fifteen mounted, thirty-two dismounted gendarmes and thirty dismounted Levies. He marched via Shaklawa, on Batas; and there is no doubt that in addition to the great difficulties of the expedition he was met by treachery on all sides. **August 30th.**

Near Batas his force was attacked, and, the whole country being now hostile, he was obliged to retreat on Arbil, being fired on from all villages passed *en route*. He only brought back thirty-one of his force out of seventy-seven. **Action at Batas.**

In November, the Political Officer, Mosul, Mr. J. H. H. Bill, I.C.S., and the Assistant Political Officer, Aqra, Captain K. R. Scott, M.C., were attacked and killed by Kurds of Zibar and Barzan near Bira Kapra. **November.** **Aqra.** **Murder of Mr. Bill and Captain Scott.**

The Kurds then attacked Aqra, which was held by Lieut. Barlow and some gendarmerie.

They put up a good fight; but had to retire. **Aqra attacked.**

The Yuzbashi (native captain) Hasoon Ibn Falayfil, who later was awarded the medal for gallantry, rallied a small party at Jujar; and by holding on here blocked the road to Mosul from the insurgents, and gave support to such Kurdish chiefs as remained loyal to the Government. He held on until relieved, and his action enabled the country up to the Aqra Dagh to be re-occupied.

In the Deir-ez-Zor area, unrest had been continuing for some time. Already two Officers, Captain Chamier, Political Officer, Deir-ez-Zor, and Lieut. Mills of the 6th L.A.M. Battery, had been ambushed on the road. **Deir-ez-Zor.**

On December the 10th matters came to a head, and the place was invested by some two thousand Arabs, supplemented by the rabble from the town. **December 10th.**

There were only sixty Levies available for defence; they were driven from the Government offices to the barracks, where there was no food or water, and made their escape by twos and threes, or became casualties until only twenty remained.

The Political Officer agreed to evacuate the town on December 11th. **December 11th.**

The Levies were given 60 mll. pay and dismissed.

Just as the year closed twenty-four men of the 4th Euphrates Levy

Samawa.

attempted to collect rifles in the Samawa area, but they met overwhelming numbers, and had to retire, losing three killed, one wounded and eight horses.

It is interesting to note that while the Levy Force was improving in strength and training, and carrying out the work described above, that a statement was made in Parliament that local Levies were not being raised and trained in Mesopotamia.

CHAPTER II

1920–1921

THIS year, 1920, opened with changes in the control and organization of the Levies. One point of contention was the Sulaimani Levy. After the defeat and capture of Sheikh Mahmud at the action of the Bazian Pass, Major E. B. Soane took over the area, and made it during his period of rule the quietest in Iraq. He was a strong character and difficult to deal with, and hence came the arrangement about the Sulaimani Levy, that Major Soane controlled its strength and use, while the Inspector-General was responsible only for equipping it. 1920. Sulaimani under Major Soane.

In March there was a new arrangement made which came into effect on April 1st. April 1st.

The force had been already, in September 1919, divided into two parts: (i) Striking Force, (ii) Police—both composed of mounted and dismounted troops. Striking Force and Police.

This Striking Force was to be considered an armed Reserve at the disposal of Political Officers, under the same conditions as laid down in the Memorandum of August 12th.

The Police were to carry out the duties hitherto done by the whole Levy Force.

By a slight change in organization the total Levy Force now was:—

24 Squadrons of 115. Total 2,760.
17½ Companies of 115. Total 2,012.

And in accordance with the above arrangement were divided as follows:—

(*a*) Striking Force: 19 Squadrons.
7½ Companies.
(*b*) Police: 4 Squadrons.
7 Companies.

One squadron and the two companies of the Sulaimani Levy were left undefined, probably some special arrangement being made between the Inspecting Officer and the Political Officer. Pay and allowances were dealt with at this time. Proficiency pay was fixed at 5 Rupees a month. Rations for men was fixed at 15 Rupees, and for horses at 25 Rupees.

Training of the mounted troops was to be definitely as mounted infantry.

The Arab Rebellion.

These arrangements had just been completed when the Arab Rebellion broke out. This was a most trying time for the Levies, who remained faithful to the Government throughout the rebellion.

They had to face the worst forms of persecution to induce them to change sides. Intensive propaganda was levelled at them by their own people, including their female relations. They were openly insulted in the streets and coffee shops, and called infidels. Reports were circulated to them that their own women, whom they had left in their homes, were being assaulted, or in some cases carried off and killed. They fully realized they were cutting themselves off from their own people.

The two indecisive actions of Mahmudiyeh and Ibn Ali were exaggerated into Arab victories. The Manchester Column disaster occurred in the middle of the country from which the 2nd Euphrates Levy were drawn. Parties of Levies who were besieged in Rowanduz, Diwaniyeh, Abu Sukhair, Kufa, Hillah, Khidr, and Nasiriyeh, were called to by name by the rebels, to come out and protect their own homes and relations.

In spite of all these trials, desertions were very few.

May 13th/14th.

Active operations began on the night of May 13/14th, when one hundred men of the 3rd Tigris Levy with one hundred Amarah and one hundred Qurnah tribesmen, made a successful night raid on the Bait Jasim and Bait Mahmud of the Nawafh, at Al Baidah in the marshes.

Raid on Al Baidah.

June 30th.

Mahmudiyeh.

On the 30th June fifty Levies mounted on horses, lent by the Sheikh of Dulaim, were attacked while reconnoitring Mahmudiyeh, by a superior force of the enemy. The horses stampeded, and the force fell back, losing five killed and eight wounded.

July 3rd–6th.

Imam Hamza.

From July 3rd–6th a body of sixty Levies accompanied a column operating about Imam Hamza, and made raids on villages round.

There were twenty-six mounted and twenty dismounted Levies holding Imam Hamza, the railway station there, and Nabi Madiyan, and these had daily skirmishes with the rebels.

July 9th.

Khan Jadwal.

On July 9th Captain Priestly-Evans and sixty Levies were attacked at Khan Jadwal. They made a successful defence; but Captain Priestly-Evans and ten men were killed and twelve were wounded. They inflicted one hundred casualties on their opponents.

July 12th.

Three days later a party of sixty Levies were surprised on the railway and dispersed.

July 14th–18th. Siege of Abu Sukhair.

From July 14th–18th a small detachment of the 2nd Euphrates Levy was besieged in Abu Sukhair. The besiegers succeeded in getting into a house, where the women and children of the Levies were, and these were only rescued with difficulty. Arrangements were made to evacuate the place, and on July 18th the besieged garrison was withdrawn to Kufa. In Kufa was a squadron of the 2nd Euphrates Levy, commanded by Lieut. F. J. McWhinnie and Lieut. Matthews. They formed part of the garrison and stood a ninety days' siege, in which they were reduced to eating the mules. The Levies lost five killed and fifteen wounded in the siege. Siege of Kufa.

In the Hillah area Levies were engaged during the whole of July and August in patrol actions with the rebels.

On August 1st three troops of Levies holding Bab el Maslakh were attacked. They were forced to retire, but did so in good order, and the enemy lost heavily. A detachment of the Indian Army and seventy men of the Euphrates Levy under Lieuts. Davies and Simpson had been holding Ain and Khidr from the 2nd July until the 12th August, assisted by an armoured train on the railway, and two boats on the river, which was too low at this time of year to allow the boats near enough to get a good target for their guns. August 1st. Bab el Maslakh. July 2nd–August 12th.

Up to August 12th the work of the Levies had been confined to patrolling; but about that date a large concentration of Arabs was reported, and they were ordered to evacuate Ain and Khidr, and go to Ur.

On the night of August 12/13th firing began, and went on all night. In the morning the armoured train moved out; but met a large body of the enemy marching on Ain village, and returned to Khidr. The town was then surrounded. The whole garrison was soon engaged, and many horses were hit. August 12th-13th. August 13th.

Another armoured train managed to get in from Ur, and the evacuation of the town began. Horses and stores were entrained, and all shunting done under heavy fire, and there was no railway official to superintend. Evacuation of Ain and Khidr.

Directly the trains began to move, the rebels swarmed down towards the station yard. Three trains in all were dispatched, the one in rear fighting a rear-guard action.

At Alu Risha one train ran into the rear of the other, and three trucks were telescoped and the line blocked. Fire was poured into the train from all round.

The rear train had to be evacuated, being the wrong side of the telescoped trucks, and all the personnel on it was transferred to the front train. Ur was eventually reached. The Levies lost eight and the Indian Army twenty, killed and missing. Fifty-nine horses were lost, and a quantity of material. Lieut. Simpson received the M.B.E. for his good work in this action. The official account of this action draws particular attention to the continuous and exhausting outpost work done by this small Levy detachment, and points out the success with which they carried it out. August 13th.

August 14th. On the 14th August the Diala Levy from Shahraban marched out, and relieved a military train held up by rebels.

August 15th. Shahraban. Next day they were attacked on all sides, ammunition ran out, and the enemy rushed Shahraban. Captain J. T. Bradfield, commanding the post, Sergeant-Major Newton and 35 Other Ranks were killed, 12 reported missing, and 15 captured.

September 10th. On the 10th September the "A" Squadron, 2nd Euphrates Levy, were attached to the 5th Cavalry for operations on the right bank of the Euphrates in the 53rd Brigade Group.

September 11th. September 12th. Actions on The Euphrates. On the 11th this column advanced on Sadr Tomaznah, burning villages as they went, and on the 12th the force was divided into two columns, the "A" Squadron, 2nd Euphrates Levy, forming advanced guard to the right column. The objective was the Khawas Canal.

Patrols of the Levies went forward at a gallop, and forestalled the rebels by a few minutes.

The next objective was Tuwairij, the Levy Squadron now becoming right flank guard.

They took the village of Beit Salman Musa at a gallop. The enemy counter-attacked, and the squadron held them off until relieved by a company of the 13th Rajputs.

As soon as they were relieved they were ordered to take a hill, 1,000 yards west of Beit Salman Musa. They galloped to the foot of the hill, dismounted, and drove the enemy off, inflicting several casualties as they went. They followed this up by a series of mounted advances to successive positions, until they reached their final objective, the Taijiyah Canal.

August and September. During August the disturbances spread to the North. This was not so much part of the Arab Rebellion in the South, which did not affect the Kurds, as a pro-Turkish agitation started in and about Arbil.

Evacuation of Rowanduz and Batas. Captain C. E. Littledale commanded the Levies in Arbil and was obliged to evacuate both Rowanduz and Batas with considerable loss. The Kurdish Levies remained absolutely loyal during this difficult period. Captain Littledale gained the Military Cross for his work.

Turning to the South again, the rebellion was gradually being crushed.

October 3rd. Badrah. On October 3rd the 2nd Tigris Levy, under command of Captain Bevan,* surrounded a rebel named Amin Beg in a fort at Badrah, and after a fight in which bombs were used, they captured him.

October 19th. October 30th. Levies were also engaged in other areas during the whole of October, and took part in successful actions against the rebels on the 19th and 30th of the month.

During November and December "A" and "C" Squadrons, 2nd

* Captain Bevan was accidentally killed later whilst on leave.

Euphrates Levy, operated in the Hillah area, and were almost continuously on the move. They were in action on 11th, 18th and 23rd November. On December 13th "C" Squadron moved to Diwaniyeh, and for the rest of that month, the whole of January and into February 1921, they and "B" Squadron of the same Regiment were engaged on operations, with columns in the Diwaniyeh area.

November 11th. November 18th. December 13th. 1921. January. February.

Throughout this period of the Rebellion the Levies lost seventy-three killed in action, and gained fifteen medals for gallantry.

Casualties of the Levies.

General A. G. Wauchope, C.M.G., C.I.E., D.S.O., left about the end of 1920, and gave on leaving a cup, still held at Levy Headquarters as a Shooting Cup. He had inspected Levies from time to time during his period of command of a Brigade in that country.

CHAPTER III

1921

1921.

EARLY in 1921 was held the Cairo Conference on Iraq, and from the decisions taken there, the future of the Iraq Levies was decided, and laid down by orders from the British Cabinet shortly as follows :—

The Cairo Conference.

" The function of the Iraq Levies, as determined at the Cairo Conference, is to relieve the British and Indian Troops in Iraq, take over out-posts in Mosul vilayet and in Kurdistan, previously held by the Imperial Garrison, and generally to fill the gap until such time as the Iraq National Army is trained to undertake these duties."

So far the Levies had consisted entirely of Arabs, Kurds, and Turkomans. Now that the Iraq Army was to be formed, the Arabs would be required to join it rather than to go to Levies.

It was decided to enlist Assyrians in the Levies.

Apart from their work already described on the Guli and Amadiyah operations they had shown good fighting qualities on the following occasions.

Attacks on Baqubah Camp and Mindan.

In September 1919, when the Assyrian Repatriation Camp at Mindan, about thirty miles north-east of Mosul, was attacked by Kurds, the Assyrians though greatly outnumbered beat off the attack, and, with a loss to themselves of four killed and eight wounded, inflicted a loss of sixty killed on their opponents, and drove them over the Zab, where many were drowned.

In July 1920, when a body of Arab rebels attacked Baqubah Camp, the Assyrians not only beat off the attack, but also took the offensive, making the attackers pay very dearly for their attempt.

Later in the year occurred the abortive attempt to repatriate the Assyrians in Hakkiari. This has been described in detail by other writers ; but, suffice to say, that a body of these people attempted to march back and occupy their country led by Agha Petros, an Assyrian leader. They went late in 1920, and the only result was a lot of fighting with the Kurds, in which the Assyrians certainly gave back something of what they had had in previous years and showed their fighting qualities. This affair was a fiasco and they returned to Iraq. This left an unsolved problem ;

but also showed that a very useful crowd of good fighting men were ready at hand with nothing to occupy them. It was therefore decided to start enlisting them for the Levies.

Enlistment of Assyrians begins.

The beginning of the enlistment of Assyrians was made at Mindan Camp. Already a memorandum had been circulated asking for officers.

On April 17th Major-General G. A. F. Sanders, C.B., C.M.G., interviewed the following Officers of the Indian Army, who had decided to undertake this work. They were Captains H. McNearnie, R. G. Ardron, J. F. Knowles, E. St. J. Hebberd and R. Merry.

April 17th.

Arrangements for pay and terms of service were arranged, and on April 19th, this party arrived at Mosul, and reported to Colonel L. F. Nalder, C.I.E., C.B.E., the Political Officer there.

April 19th.

They were joined by Dr. W. A. Wigram there, and on the 20th the whole party went to Mindan Camp. However, the attempts to enlist men were not successful. Many recruiting meetings were held, but the men would not come forward. The leading men were interviewed, but only made objections. All they wanted was that the British should send them back to their country, which they had lost through joining the Allies.

April 20th. Enlistment of Assyrians.

After a great expenditure of eloquence by Dr. Wigram, and a good deal of action by the British Officers, some fifty men were got together and they, at the last minute, tried to get out of it, but were stopped.

The British personnel decided to hold on to what they had, and to prevent the men they had getting away to the camp again ; Captain Hebberd with ten men and their families went off to Aqra ; Captains Merry and Ardron with the rest went to a camp at Nebi Yunis, close to Mosul.

Captains McNearnie and Knowles stayed at Mindan to recruit and forward men to Aqra and Mosul. Colonel Nalder at Mosul fitted the force out with tents, and through him the ration contracts were made. Clothing, arms and S.A.A. were drawn from the Police.

Early in June, by recruiting locally and by drafts from Mindan, the force rose to about 250 at Mosul, and was organized as No. 8 and No. 9 Companies.

June.

The British Officers selected and appointed the Native Officers and N.C.O.'s, and of these many proved to be failures and were broken at once. All those picked to be officers, in any case did a period as N.C.O.'s. The first Native Officer appointed was Rab-Khamshi Yusuf Yokhana, who became Signalling Officer of the 1st Assyrian Battalion ; and on about the same date, Daniel Ismail, son of Malik Ismail of the Upper Tiari Assyrians, was made an officer at Aqra, and is now Rab-Tremma of the 2nd Assyrian Battalion. The next two chosen were Yakub Ismail, another son of Malik Ismail, and

The first Assyrian Officers.

Shimoel Tiya, the former became Rab-Tremma, the latter Rab-Emma in the 1st Assyrian Battalion. Both were made in Mosul.

End of May.

At the end of May No. 8 Company under Captain Ardron, and No. 9 Company under Captain Merry left Mosul for Dohuk. They were escorted by two troops of the 7th Hussars. This, their first appearance on the

March to Dohuk.

march, was not impressive. They were badly equipped and clothed, and with their families hanging on to them, made a poor show. They remained at Dohuk for some three months. Another Company No. 7 under Captain W. H. Crawford Clarke, M.C., joined them there. The situation being

August 22nd.

on August 22nd No. 5 Company (Hebberd) and No. 6 Company (Moody) at Aqra and No. 7 Company (Crawford Clarke), No. 8 Company (Baddiley *vice* Ardron) and No. 9 Company (Merry) at Dohuk. All these companies were Assyrians except No. 7, which was half Kurd and half Assyrian.

In July Colonel-Commandant Sanders left the country, handing over the Levy administration to Colonel-Commandant Frith.

Lieut.-Colonel C. R. Barke, C.B.E., T.D., took over command of the

August 20th.

Levies in the Mosul area on August 20th, from Captain H. D. McNearnie, to whom is due much of the credit for the successful raising and organization of the Assyrian Levies.

At this date the Headquarters 18th Indian Division (General Fraser) and 54th Brigade (General Nightingale) were still in Mosul. The A.D.M.S. of this Division dealt with the Levies in medical and sanitary matters.

August 24th.
August 27th.

Lieut.-Colonel Barke carried out inspections of the companies at Dohuk on August 24th and at Aqra on August 27th.

August 22nd.

On August 22nd the 5th Levy Cavalry Regiment came to Mosul; this was preparatory to taking over duties hitherto carried out by the Indian Army.

September 10th.

On September 10th the relief of the Imperial Garrisons began. Two squadrons, 5th Levy Cavalry Regiment, under Captain C. O. L. Devenish,

Relief of Imperial Troops.
September 12th.

left Mosul for Dohuk; the rest of the Regiment left Mosul on September 12th and went to Tel Afar, where they took over from the Imperial Garrison there, which was one squadron 8th Hussars, one section R.F.A. and one company 3/70th Burma Rifles.

September 11th–14th.

Arrangements were rather interfered with by trouble with the Surchi Kurds, and from September 11th–14th No. 5 Company were away from Aqra at Kelaiti, co-operating with the gendarmerie, who were operating

Dasht-i-Harir.

against Sheikh Obedullah of Bajeel, in the Dasht-i-Harir. The Royal Air Force took action also. The Levies had no fighting and returned on September 14th.

September 18th.

The next relief of Imperial troops was at Zakho. Nos. 8 and 9 Companies left Dohuk and took over there from the 3/70th Burma Rifles.

Captain Merry became O.C. Station with Lieut. P. J. T. Baddiley and Captain A. C. Prevett, Officers Commanding Companies.

The mixed No. 7 Company, half Assyrian and half Kurd, and other Moslem races proved unsatisfactory. There had been trouble on the road up, and at Dohuk there was still more. After investigation the company was broken up, the Assyrians were sent to Aqra, and Captain Crawford Clarke took the rest off by kellek * from Mosul for Baghdad in November. September 20th.

In the middle of September Brigadier-General L. W. de V. Sadleir-Jackson, C.B., C.M.G., D.S.O., took over the command of the Levies, from Colonel Commandant Frith. Orders regarding training were issued, and instructors were to be detailed from the Army. Lieut.-Colonel Barke started equitation courses under the 30th Lancers, and drill and musketry courses were commenced under the 2nd Bn. East Yorkshire Regiment during October; these were all going well by the end of the month. September. Brig.-General Sadleir-Jackson, in command.

During the month the Levies received in equipment, one thousand rifles from Baghdad, and winter clothing. September.

Medical arrangements also began to take shape. Captain J. W. Malcolm, O.B.E., M.C., R.A.M.C., and Assistant-Surgeon C. L. Smith arrived at Mosul, to start Levy medical arrangements, although for some time the Director of Health Services (Civil) had the responsibility of Levy health matters.

The strength of the force on October 1st had been 21 Squadrons, 10 Companies and 2 Batteries. The Inspector-General decided on a complete reorganization, which had already been partly put in train on paper; therefore from a somewhat disorganized force of odd units the force became 4 Cavalry Regiments, 1 Pack Battery, 2 Battalions Infantry, and 3 Machine-Gun Companies. October. Reorganization into units.

General Sadleir-Jackson decided also to equip the force with automatic weapons, and Lewis and Hotchkiss guns were ordered.

Although Levy Headquarters were in Baghdad, the Levy details Camp at Nineveh still received and checked stores, and passed personnel through. Sergeant-Major Higgins remained in command.

The 18th Division left Mosul on October 15th. October 15th.

The location and strength of the Levies at the end of 1921 was:—

Mosul Area.

Zakho	2 Companies 2nd Battalion.
Aqra	2 Companies 2nd ,,
Dohuk	H.Q. 5th Regiment (less 2 Squadrons). H.Q. 2nd Battalion (less 4 Companies).
Tel Afar	2 Squadrons 5th Regiment.

* A raft supported on inflated skins.

SOUTH KURDISTAN.

Arbil . . .	4th Regiment.
Kirkuk . . .	2nd Regiment.
Sulaimani . .	One Squadron Sulaimani Levy.
	One Company Sulaimani (Infantry).
Rania . . .	One Platoon ,, ,,
Chemchemal . .	½ Company ,, ,,
Halebja . . .	One Platoon ,, ,,
Khanaqin . .	3rd Regiment.

EUPHRATES.

Diwaniyeh	1st Regiment (less 1 Troop).
Samawah	1st Battalion (less 2 Companies and 2 Platoons).
Rumaithah	1 Platoon 1st Battalion.
Nasiriyeh	2 Companies 1st Battalion.
Baghdad	1 Troop 1st Regiment.
	1 Platoon 1st Battalion.

Total strength was:—Mounted Troops 2,203.
Infantry 2,051.

Since early in the year, the Levies had not been actively engaged except for a very small affair of shooting at Dohuk, in which the 7th Company were involved, on the night of October 4th. *(October 4th.)*

December 14th. *Action at Babachikchek.*

On December 14th, however, the Arab Levy Cavalry under Lieut. H. E. Bois, acting as escort to the Assistant Political Officer, were attacked by Kurds near Babachikchek. Lieut. Bois was wounded, and nine men and horses were killed, and the force had to retreat.

Dasht-i-Harir operations.

General Sadleir-Jackson issued orders for operations against the Kurds on the 16th December; a Cavalry Column some 7–800 strong to concentrate at Arbil under his own command, and to march via Shaklawa on Batas, and an Infantry Column from Aqra under Lieut.-Colonel Barke, to march on Batas via Kandil, and co-operate with the cavalry. The concentration produced some hard marching. A squadron of the 5th Levy Cavalry Regiment, for instance, left Dohuk on December 18th and picking up another squadron under Captain J. P. Carvosso of the same Regiment, which marched from Tel Afar, arrived at Arbil on December 22nd, doing 110 miles in all.

December 25th.

At 18.00 on December 25th the Cavalry Column left Arbil under General Sadleir-Jackson, Captain Devenish following at 06.00 next day with the ration convoy.

The Cavalry Column reached Sisawa on December 26th, and found the Kurds in occupation of Harir and Batas villages. December 26th.

To return to the column under Lieut.-Colonel Barke, which advanced from Aqra. He had sent forward one platoon, and with it an Arab Kelekchi, with the necessary skins and poles for making kelleks. December 24th.

Captain McNearnie was in command. This advanced party reached Isteria village on December 24th, and began the construction of the raft. They came under fire at once from the opposite bank. Action at Kandil.

The rest of the Infantry Column under Lieut.-Colonel Barke was at Bajil on December 24th. At 15.00 on December 25th, they reached Isteria, and found McNearnie's party engaged with Kurds on the opposite bank, and all raft-making stopped by the fire. December 25th.

Rab-Khamshi Polus Elias had been wounded.

During the night the raft was completed. At 06.30 on December 26th the first raft load of fifty men was sent across covered by rifle fire and the fire of one machine-gun worked by Sergeant-Major Hillier, who had just brought it up to the Mosul area for instructional purposes. The party came under heavy fire in crossing, but succeeded, and the enemy force, which contained some Turkish soldiers from Rowanduz, retired. The crossing was completed at 11.00. At 11.30 two aeroplanes of No. 55 Squadron appeared. One landed and crashed in trying to take off again. The pilot was taken off by the other, while the mechanic of the crashed aeroplane with his Lewis gun, joined the column. Barke's Column halted at Khorra near Kandil. On December 27th they burned Khorra, and advanced on Batas, arriving at a point one mile east of it at 11.05. December 26th. December 27th.

Meantime the Cavalry Column advanced from Sisawa, and attacked Harir. Here they met a determined resistance, and by about 11.00 they were definitely held up. The fight went on all day and by the evening Captain Carvosso and five of his men had been killed. Lieut. R. A. Burridge was mortally wounded next morning. Action at Batas.

No orders came to Barke, who then attacked Batas. This was done at 12.45 with air support, and at 13.40 the village was occupied by Captain E. St. J. Hebberd with the loss of two men wounded. Very heavy rain now began. The Cavalry Column was still about Sisawa, with two squadrons holding a strong outpost line near Harir.

At 17.00, Captain Devenish, who had just arrived with the convoy, was ordered to march with the convoy escort, as soon as the men had had a meal, to attack the Harir Dagh under cover of darkness.

He left his horses at Sisawa, and marched at 17.30, got partly up the Dagh in the dark, and then became stuck in the rain and darkness.

On December 28th the action was renewed. Devenish moved on up December 28th.

the Dagh, as soon as he could see, captured one Kurd on the way up, and near the top came into action. A squadron of the 3rd Regiment followed him, and the Kurds withdrew. He reached the top at 08.00, pushed along the Harir Dagh, to a point above Batas, and here took up a position and built sangars. Several small parties of enemy were seen, but did not come near, and made off when fired at.

Action at Batas.

In the morning Barke's Column started from Batas to join the Cavalry Column at Harir. The rear-guard was hotly attacked as it left by Kurds pressing on through the gardens. R.T. Daniel Ismail commanding the rear-guard, counter-attacked, bringing back seven rifles and one sword. At 15.00 General Sadleir-Jackson, finding it impossible to obtain air co-operation owing to the inclement weather, ordered Barke to attack Harir. This was done in co-operation with Captain Littledale and the Police with a loss of two killed, one died of wounds, and one wounded. The town was burnt.

This ended the operations.

December 29th.

On the morning of the 29th December Devenish, who had spent the night in the snow on top of the Harir Dagh, sent patrols out but found nothing. The columns all concentrated at Sisawa, and left for Shaklawa. Captain Devenish and Lieut. D. S. Foster followed with their party from Harir, half a company of Assyrians from Barke's Column holding Sisawa until they arrived.

December 31st.

The whole force was back at Arbil by December 31st.

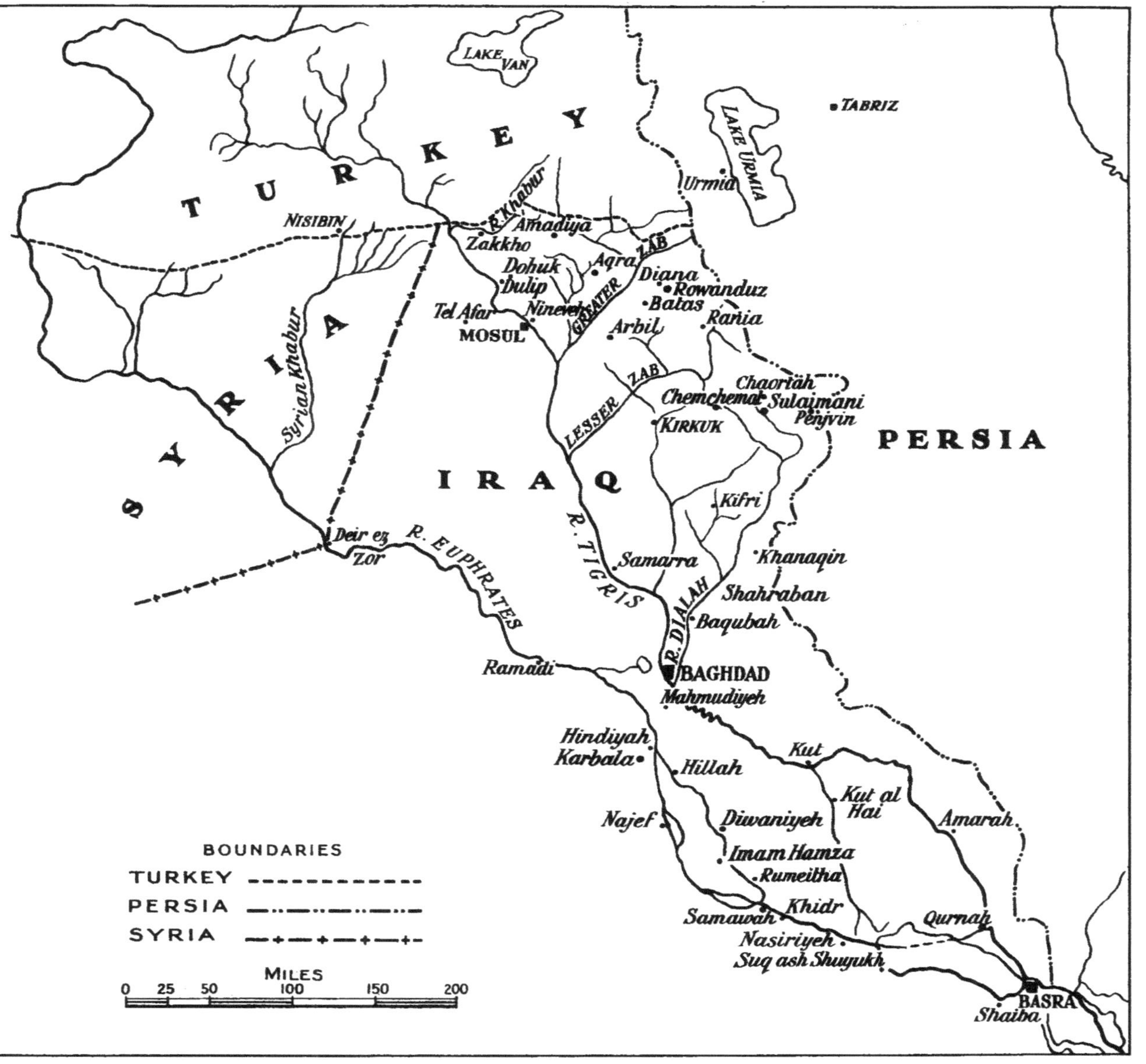

Areas of Employment of Levies from 1915-1921.

CHAPTER IV

1922

ON 17th January, 1922, the Levies were placed under G. O. C.-in-C. Military Forces, except for finance and administration. 1922. January 17th.

The strength and organization of the force was fixed, after discussion, at :—

Levies under G.H.Q.

3 Regiments of Cavalry.
4 Battalions of Infantry.
1 Machine-Gun Company.
2 Pack Batteries.

The Cavalry had no automatic weapons. The Infantry Battalions did not get Lewis guns until July. Up to this date there had been four Regiments of Cavalry, the reduction from five to four having been just effected. Now another had to go, and to bring this about the 2nd and 3rd Cavalry Regiments were amalgamated, and re-named the 3rd Cavalry Regiment, while the 4th became the 2nd Cavalry Regiment.

In May 1922, the strength of the force was :—

Cavalry, 1,410—a Regiment was 457.
Infantry, 3,248—a Battalion was 701.
Battery, 210.
Depot, 173.

Vickers guns were at this time held with Battalions ; but a change was being effected, and a Machine-Gun Company formed.

In budgetting for this force, the yearly cost of one squadron was 157,728 Rupees and of a Company 170,928 Rupees.

After the action at Batas, General Sadleir-Jackson ordered the enlistment of 1,500 more Assyrians for the Levies. This enlistment began on January 7th.

No. 6 Company (Captain Moody) went from Aqra to Dohuk to form a nucleus, and to assist in training new recruits. Captain McWhinnie was transferred from the Euphrates area to Aqra, to raise an Assyrian Company there. Lieut.-Colonel Barke proceeded to Dohuk at once and the enlistment of Assyrians was carried out, partly by a party consisting of Lieut.-Colonel Barke, David D' Mar Shimun, the father of the Patriarch, and Rab-Emma Daniel Ismail, son of Malik Ismail, and the senior Assyrian Officer of the Levies, and partly by Captain McNearnie, who had previously been so successful in raising the original companies, making extensive recruiting tours through Kurdistan, particularly the Amadiya area and north of it, where there were many settled Assyrians. The first party interviewed Mr. R. F. Jardine, the Administrative Inspector, and the rest of the Assyrian Maliks, but the results were not entirely satisfactory. One particular case being that of the Bohtanis, who were asked for men; the men selected sold their crops and came down to enlist and were then told they were not required.

Further enlistment of Assyrians.

February 1st. Pack Battery formed.

Fifty men of the Assyrian Companies at Zakho and Aqra were also selected for training for the Pack Battery. Guns were obtained from the Army at Mosul and on February 1st Captain Devenish left with this party by kellek, from Mosul to Baghdad.

February 1st. Machine-Gun Company formed.

A further party of one Assyrian Officer, and fifty Other Ranks, was sent from Mosul to Kirkuk, to be trained as machine-gunners at the Machine-Gun School there under Lieut. Simpson, assisted by a staff of B.N.C.O.'s.

January 24th.

Courses were also arranged in (*a*) Bugling, eight men, (*b*) Kellek making, ten men; this was very necessary for operations in the hills, and (*c*) Carrier Pigeon Work, one man.

Training.

Eight instructors, four in drill and four in musketry, were sent from the 2nd Bn. East Yorkshire Regiment to the camp at Dohuk.

February 26th.

On February 26th Lieut. J. B. F. Austin, 7th Hussars, arrived at Mosul with W/T personnel. He took over a pack W/T set from the R.A.F. and proceeded to Dohuk.

March.

By the beginning of March training in the camp at Dohuk was going well. A report, of that date, says that the musketry of the Levies was better than their drill.

In spite of certain set-backs, recruits were coming in well. McNearnie's recruiting trip had been very successful, and by the middle of March some twelve hundred recruits and families were assembled at Dohuk.

Captain Prevett was at this time in command of Dohuk, with Captains B. C. Moody and H. A. Foweraker as Officers Commanding Companies. As the present site was very cramped and within the range of snipers from the adjacent hills, Dohuk Camp was moved to Dulip at the end of March.

March 14th.

On March 14th General Sadleir-Jackson handed over the command of the Levies to Colonel G. R. Frith. A further re-organization of command took place in April. Levy areas were abolished. Mosul Area ceased to exist on April 12th, and the next day Lieut.-Colonel Barke went to Dulip, and took over command of the battalion there. He had under him Captains Prevett, Moody, Young, Baddiley, and Foweraker. The latter was in command of a new company which was forming. There were also eight British N.C.O. instructors from the East Yorkshire Regiment, one N.C.O. sent by Mosul District for physical training and one W/T set. In addition there were twelve hundred recruits and fifteen hundred women and children in the camp.

April 12th.

Dulip Camp.

The units under his command in Dulip Camp were 2nd Battalion Headquarters and two companies, the other companies being at Zakho.

Training and equipment.

The 3rd Battalion with two companies at Dulip and two at Aqra. At this time battalions had companies of 240 but no headquarters wing. The 2nd Cavalry Regiment was also at Dulip. The latter consisted of three squadrons of Kurds and was under the command of Major E. N. Eveleigh, D.S.O., M.C. It was decided to form an Assyrian Squadron in this Regiment, and Lieut.-Colonel Barke transferred a certain number of his men to the Regiment for that purpose.

This was done partly to take the place of the Arab personnel of the Levies, as all Arabs in the country were now required for the Iraq Army. Orders were issued that no more Arabs were to be enlisted, and that those already in the Levies were not to be taken on as their time expired.

Barke commanded both battalions at Dulip until the arrival of Major Bentinck. Prevett was appointed adjutant of the 3rd Battalion and Moody acted as adjutant of the 2nd Battalion.

It will be readily understood, especially by those who remember the initial stages of the raising of the New Army in 1914, that the raising and formation of these battalions was not accomplished without difficulties being met with and the occurrence of the occasional hitch.

Whilst a certain number of Assyrian N.C.O.'s were sent from the companies which were raised nine months previously, the bulk of the A.O.'s and N.C.O.'s had to be selected and appointed from the twelve hundred recruits recently joined. At that time there were few of them with much knowledge of English and the B.O.'s and B.N.C.O. instructors had no knowledge of Syriac. In spite of all difficulties the British Officers and N.C.O.'s worked hard at the training and General Nightingale visited the camp from time to time and encouraged everyone's efforts.

There also occurred a delay in the issue of clothing, equipment and tentage; a considerable portion had been obtained from the Army Ordnance Depot at Mosul, just before it closed down in January; the remainder had to be obtained from Baghdad and through some misunderstanding was very late in arrival.

In March the recruits in camp at Dohuk, numbering between two and three thousand, could not be clothed and equipped immediately, became restive and desired to go back. Barke had to address them and assure them that he had seen their clothing in large boxes being loaded on to camels at rail-head (Shergat) and that it would arrive shortly.

June.

In June the 2nd Cavalry Regiment came down from Dulip to Nineveh Camp, the building of which had been completed by Captain W. E. Parnell. The Squadron of Assyrians was formed, under Rab-Emma Shain Gewergis, and put under training.

Malaria at Dulip.

Meantime Dulip Camp had proved most unhealthy. Malaria began at the end of May and by August practically every British Officer and British N.C.O. had been attacked and also some seventy per cent. of the Assyrians. The G.O.C. (General Fraser) visited the camp and inspected all the troops and issued orders for the 3rd Battalion (Barke's), less the two companies at Aqra, to move to Mindan. This was due to an attack being expected on Erbil by the Kurds and Turks. The column left Dulip on 13th July and reached Mindan on the 16th. Several of the party could not march owing to malaria. There were no water-bottles, and water had to be carried in chagals, purchased out of P.R.I. funds. One hundred and ten camels had to be hired to convey baggage and S.A.A. On arrival at Mindan, signallers lent by the 11th Rajputs established communication with Mosul from the top of Jebel Maglub.

After G.H.Q. had taken over the Levies, as a result of a conference, the pay of 50 Rupees a month promised to recruits by McNearnie, as he was authorized to do on his recruiting tour, was changed to 45 Rupees a month. This was, in effect, a breach of faith. The only thing the Battalion Commanders could do was to call up the Assyrian Officers, tell them of the decision of G.H.Q., and tell them to inform the men.

As it happened, nothing went wrong, and the men accepted the situation without comment.

But it was realized that by 1923 owing to the refusal of the Assyrians to re-engage that the Assyrian units would cease to exist by the spring of that year. A conference was held at Dohuk under the Colonel-Commandant in the autumn of 1922 when more favourable terms were offered and the Maliks asked to co-operate. At first an indecisive situation was the result, but finally the Patriarchal family threw their influence into the raising of recruits and re-enlistments. David D' Mar Shimun, father of the Patriarch, became an Officer in the Levies and as a result of his efforts, together with those of the other recruiting parties, practically all serving

Dohuk Conference with Assyrian Maliks on Recruiting.

Assyrians re-enlisted and a sufficient number of new recruits was raised to save the whole situation. Since then no trouble has ever been found to raise recruits.

In January the first contingent of the Iraq Army arrived in Mosul. They were played in by the band of the East Yorkshire Regiment. Another battalion of the Indian Army, the 15th Sikhs, left at the same time. Arrangements were made at once for the Iraq Army to take over Tel Afar from the Levies.

Iraq Army arrive at Mosul.

June.

In June, General Nightingale left. His departure was much regretted by all the Levies, whom he had helped greatly. He had had contingents out from time to time, and reported well on them.

June 24th.

On July 7th the old Euphrates Levy Head Quarters closed down; they had been on the right bank of the Tigris at Baghdad, in the lines occupied later by the 1st Marsh Arab Battalion of the Levies.

July 7th.

In October the Royal Air Force took over from the Army. This had been decided on at the Cairo Conference, and as this forms a definite landmark in the history of the Levies, as well as of Iraq, the operations they were engaged in from the beginning of 1922 to this date will now be described.

Royal Air Force takes over in Iraq.

In January the Sulaimani Levy carried out operations against the Avroman Kurds, helped by some of the Jaf near Halebja. In action at Kurmal, Captain H. C. D. FitzGibbon, 13th Hussars, serving with the Levies, was killed in action.

January 16th.

Avroman operations.

In the Northern Area matters were quiet at the start of the year, though there was some anxiety about the Kurds attacking Aqra in February. This came to nothing.

February.

In May and June, General Nightingale carried out operations against the Kurds in the North-Eastern Area. At Baneh Banok, north of Halebja, an action was fought in which Lieut. V. T. Mott and four Other Ranks of the Levies were killed.

June 1st.

These operations were hardly over when trouble began in the Chemchemal area.

Already at Jabbari on May 21st, the Mudir had been attacked and wounded by Sayid Muhammad Jabbari and his escort deprived of horses and arms.

May 21st.

The Begzada Section of the Hamawand were restive at the curtailment of their privileges; Kerim Futteh Beg, their leading spirit, openly threatened rebellion.

Trouble with the Hamawand Kurds begins.

Propaganda for the return of Sheikh Mahmud was continuous, and having its effect.

On June 12th the Political Officer asked for a Levy Force to proceed

June 12th.

June 16th. to Tainal. On June 16th it arrived, consisting of the Sulaimani Levy Battalion of three unmounted and one mounted companies, and one section of Assyrian machine-gunners, from the Machine-Gun School at Kirkuk, all commanded by Lieut.-Colonel E. C. T. Minet, D.S.O., M.C. (Reserve of Officers), Iraq Levies. June 17th. On June 17th the principal headmen were summoned and interviewed by Captain S. S. Bond, the Assistant Political Officer.

Kerim Futteh Beg then wrote professing loyalty, asked why the military had come, and offered to meet Captain Bond at Mortaka.

June 18th. On June 18th Captain Bond, accompanied by Captain R. K. Makant, who was about to transfer from the Levies to the Political side, and was to succeed Captain Bond at Chemchemal, proceeded to Mortaka. They were met by Kerim Futteh Beg two miles from the village. As they rode in with him both officers were shot in the back, it is said by Saber and Abdullah, sons of Kerim Futteh Beg, and killed instantly.

Murder of Captains Bond and Makant.

Chase after Kerim Futteh Beg.

Colonel Minet recovered the bodies and buried them the same day and at once took action to hunt down Kerim Futteh Beg and his band.

The 1st Levy Cavalry Regiment joined him from Kirkuk, its place being taken by a squadron of the 3rd Levy Cavalry Regiment, and for the next month, Minet carried out a vigorous hunt after Kerim Futteh Beg, beginning in the Chemchemal area, thence to the Diala River in Sangaw, and from there northwards to the Surdash valley. Another squadron of the 3rd Levy Cavalry moved up from Khanaqin to Maidan on the 20th to assist in the operations. July 20th.

July 23rd. On July 23rd came news, premature as it happened, that Kerim Futteh Beg had crossed the Lesser Zab at Dukhan. Minet started in pursuit at once and arrived at Derbend on July 27th, the Headquarters of the Assistant Political Officer, Rania, Mr. Edmonds.

Further trouble had begun in this area. The Pizhdar tribe had been for some time in a state of disaffection. June 23rd. Moreover on June 23rd a Turkish officer named Euz Demir, with a party of officers, had arrived in Rowanduz, with a special mission of stirring up tribal revolt by Pan-Islamic propaganda, and presents of ammunition and the like.

Turkish action in Iraq.

July 10th. July 11th. July 18th. By July 10th the situation was such in Rowanduz, that it was bombed by a concentration of aeroplanes on July 10th, 11th, 18th and at intervals later.

Rowanduz bombed.

July 29th. On July 29th Colonel Minet moved his force to Qala Diza, accompanied by the Assistant Political Officer. No action was sanctioned against the Pizhdar, who were in their summer-camps in Persian territory, but this move had good effect generally.

Pizhdar unrest.

August 1st–4th. They remained in Qala Diza from August 1st to 4th. It was unhealthy, full of mosquitoes, and there was a severe heat-wave. Eighty per cent of the force went down with malaria. They came back to Derbend on August 5th, stayed two days, left for Sulaimani via Koi Sanjak on August 9th. Captain C. E. Simpson, M.B.E., I.A.R.O., serving in the Levies, was evacuated by air to Sulaimani; but died on arrival there. August 5th. August 9th.

Colonel Minet left one hundred rifles of the Sulaimani Levy and the Assyrian Machine-Gun Section at Derbend, under Captain J. Bourne, R. of O., 4th Battalion Levies. He left Captain F. J. McWhinnie, R. of O., with one company Sulaimani Levy at Chemchemal, and went on to Sulaimani.

Kerim Futteh Beg had gone through Koi Sanjak area and reached Rowanduz on August 9th. August 9th.

On August 16th the Turks advanced, reached Nawdesht that day, and the Shawr valley, north of Rania, on August 18th. August 16th. August 18th.

Tribal Lashkars were collected to hold up the Turkish advance and defend Rania and villages occupied by Turks were bombed. The bombing was chiefly done by Flight-Lieuts. Robb and Kincaid, late the Schneider Cup winner.

On the 18th August G.H.Q. ordered the formation of Ranicol. This consisted of: Ranicol formed.

Two Companies and One Machine-Gun Platoon, 15th Sikhs.
One Section, Ambala Pack Battery.
One Squadron, 1st Levy Cavalry Regiment.
50 Men, 3rd Levy Cavalry Regiment.
4th Battalion Sulaimani Levy.
One Machine-Gun Platoon, Assyrian Levies.

Part of this force, namely, one hundred men of the Sulaimani Levy and one section of the Assyrian Machine-Gun Platoon, were already there. One company of the 15th Sikhs and the fifty men of the 3rd Levy Cavalry never went beyond Koi Sanjak.

The first units of this column arrived on August 29th. By the time these troops reached Derbend, the Merga and Chinara Nahiyahs of Rania and Qala Diza were all out of control. August 29th.

On August 30th one dismounted and one mounted company of the Sulaimani Levy, under Captain H. E. D. Orr-Ewing, held Rania; the rest of the force was at Derbend. August 30th.

Opposite Derbend, across the Lesser Zab on its south side is a high rocky ridge. This was held by a piquet of twelve men of the 15th Sikhs. At dawn on August 31st, the Kurds rushed this piquet, killing ten, while two wounded got away. August 31st. Pizhdar attack Derbend.

From the ridge fire was opened on Derbend Camp and kept up, in spite of air attacks. Many casualties occurred among the horses and mules.

As soon as the piquet was lost, orders were sent to Captain Orr-Ewing

August 31st. to march to Derbend. These orders were countermanded; but directly he left Rania it was occupied by enemy parties. At dusk, Captain Orr-Ewing bivouaced at Boskin.

Retreat from Derbend. Colonel Hughes, 15th Sikhs, commanding at Derbend, ordered the evacuation of that place at dusk.

Near Kurrago village the column was fired into and thrown into confusion. It was re-formed and at Boskin finally joined Captain Orr-Ewing, who came there from Rania. Boskin is a village with a mound, trees and plenty of water.

Now that Rania was held by the enemy, Colonel Hughes decided to

September 1st. retire to Koi Sanjak. The march began at 10 a.m. on September 1st. The plain was full of tribesmen who attacked the column from all sides. The transport, in a panic, got in among the advanced guard. The Sulai-

Disaster to Ranicol. mani Cavalry, whose officer two days later deserted to the enemy, took no part in the action. The company of Sulaimani Levy which with the company of 15th Sikhs formed the rear-guard and the Assyrian machine-gunners did well.

The column had been marching straight across country, and one part of the Rania plain is a big area of rice-fields. Into this area the column now came, transport and advanced guard mixed up together. Some of the transport became stuck, one gun and a lot of baggage was captured by the Kurds. Here a bayonet charge by the 15th Sikhs accounted for several of the enemy. Other Kurds from the north were coming down on the column from that flank and then just as the column was emerging from the rice-fields two aeroplanes arrived from Kirkuk and machine-gunned the enemy, especially those coming down from the north. They now ceased to press the column which reached Gird-i-Buraise that day

September 2nd. and Koi Sanjak on the 2nd September. Total losses of the column were 27 killed, 7 missing, 32 wounded, of which Levies had 13 killed and 19 wounded.

The Assyrian machine-gunners, who had done well, were thanked at the end of the operation by the General Officer Commanding.

September 6th. O/2265/328 of 6/9/22. The result of this regrettable affair was that matters became so difficult in the whole eastern area that evacuation of Sulaimani was decided on.

Evacuation of Sulaimani. A mixed population of sixty-five was evacuated by air, including the Levy officers there. Sheikh Mahmud was left as president of a local elective council to run Sulaimani, and the officers and men of the Levies, with arms and equipment, were handed over to Sheikh Mahmud. The personnel were struck off Levies, but pay continued to be issued to them through the Levy budget. This ceased on Sheikh Mahmud objecting to the arrangement. Some of the personnel of the Sulaimani Levy were on leave, and some in Kirkuk. A certain number tried to rejoin the Levies. This was discouraged, so that Sheikh Mahmud should not be able to say inducements were being held out to the men to leave him.

Captain McWhinnie's detachment at Chemchemal knew nothing of the evacuation, until aeroplanes passing over, with personnel who were being evacuated, dropped a message on his camp. He remained for the present at Chemchemal.

Another affair with unsatisfactory results occurred in the Amadiya area.

A party of Barzan Kurds under Sherif Agha, had got into the town, and only the timely arrival of a party of Assyrians under the Metropolitan Bishop Mar Sergius, saved the place from looting and destruction and the life of the Qaimakhan. **September.**

Orders were issued for action to be taken against the Barzan Kurds, and a column consisting of part of the 2nd Battalion Levies and a body of Irregular Assyrians, under Zia D'Mar Shimun of the Patriarchal House, and Malik Ismail of the Upper Tiari, the whole being under the command of Lieut.-Colonel Bentinck, advanced on Barzan from the west. **Barzan operation.**

Another small column of two companies of the 3rd Battalion and the Levy Pack Battery, under command of Major Lake-Geer, marched to Aqra.

The Irregulars, with Bentinck's column, had pushed on and occupied Barzan, and he with the battalion was at Belinda, when orders came to him from G.H.Q. to return at once. He did so, by a night march from Belinda. No news of this seems to have reached the Irregulars, who after holding Barzan for a short time against the attacks of the Kurds, made their way back with some casualties and great difficulty.

The other small column had occupied the top of the Aqra Dagh and one company had burnt Naqaba village in the valley below, when the same orders came, and they went back also.

The affair had the unfortunate effect locally of making the Barzan and Zibar Kurds imagine that the column had retired on account of them, and this gave them quite a false impression of their own power.

CHAPTER V

1922–1924

1922. October. Royal Air Force takes over from G.H.Q.

IN October came the change, when the Royal Air Force took over from the Army. Levies, in coming under command of the Air Vice-Marshal, had the following chain of communication. From O.C. Levies to the Air Vice-Marshal, thence to the Air Ministry, and from there on to the Colonial Office.

Colonel H. T. Dobbin, D.S.O., takes over Levies.

In October also, Colonel H. T. Dobbin, D.S.O., took over command of the Levies, with the rank of Colonel-Commandant. The late re-organizations and moves had left the Levy units very scattered. They were as follows :—

Levy Head Quarters	Baghdad.	Moved to Mosul at the end of 1922.
2nd Cavalry Regiment	Mosul.	
Pack Battery.	Mosul.	
3rd Battalion (less 2 Companies).	Mosul.	
4th Battalion.	Mosul.	Partly formed.
Ordnance.	Mosul.	
M.G. Company Head Quarters.	Mosul.	Machine Guns were brigaded.
2nd Battalion (less 3 Companies).	Dulip.	
2 Companies 2nd Battalion.	Zakho.	
1 Company 2nd Battalion.	Feishkhabur.	
2 Companies 3rd Battalion.	Aqra.	
3rd Cavalry Regiment.	Arbil.	
1st Cavalry Regiment (less 1 Squadron).	Kirkuk.	
1st Battalion.	Nasiriyeh.	
1 Squadron 1st Cavalry Regiment.	Khanaqin.	

And of these, to show how races were represented :—

1st and 3rd Cavalry Regiments were Kurds and Turkomans.
2nd Cavalry Regiment. Kurds and Assyrians.

1st Battalion. Marsh Arabs.
2nd and 3rd Battalions. Assyrians.
4th Battalion. Kurds (one Company Mounted).
Pack Battery. Assyrians.

A little later Yezidis were enlisted.

A proposal was made in November to change the name Levies once again, to Iraq Frontier Force. It was not agreed to. November.

During all this period, there was great anxiety about Turkish action against Iraq. In March there was a scare in Zakho, on account of the arrival of a Turkish general and staff at Jezireh Ibn Omar, and during the year arrangements were made to meet an attack, in which Levies formed part of two columns, the one operating from Mosul, and taking up a series of defensive positions, the other to be in the area Mindan-Arbil-Kirkuk. Evacuation of families was arranged. March. Anxiety about Turkish action.

A small body of Turks under Euz Demir were holding Rowanduz, and by communicating with Sheikh Mahmud, who in November proclaimed himself King of Kurdistan, and disseminating propaganda among the Kurds, were making matters more and more difficult.

In February, Lady Surma D'Beit Mar Shimun, the aunt of the Patriarch, wrote in from Bebadi to say that the Tiari and Tkhuma Assyrians, near the Turkish Border, were in danger of attack. 1923.

It was then decided to occupy Rowanduz, and clear up the whole situation there.

Already one small local operation had taken place. On January 19th, the 3rd Regiment Iraq Levies and Police marched from Arbil, and on January 23rd with air operation occupied Wanka without opposition. On the night of the 23rd, however, a Levy post was attacked. Two men were killed and a machine-gun captured, but subsequently recovered. January 19th. Affair at Wanka.

In March the Rowanduz operations began. The force was in two columns, Koicol consisting mainly of British troops under command of Colonel-Commandant B. Vincent, C.B., C.M.G., and Frontiercol consisting entirely of Levies under command of Colonel H. T. Dobbin, D.S.O., Colonel-Commandant of the Iraq Levies. The Air Vice-Marshal Sir John Salmond, K.C.B., C.M.G., C.V.O., D.S.O., commanded the whole operations. Rowanduz operations.

The scheme of operations was an assembly of Koicol at Koi Sanjak, and Frontiercol at Arbil, and then advance on Rowanduz. Koicol coming in from the south via Baliassan and Frontiercol from the west via the Spilik Dagh and Rowanduz gorge. Both columns were closely supported by air action.

On March 26th Lieut.-Colonel Minet marched from Mosul with the March 26th.

4th Battalion Levies, the Pack Battery and two platoons of Vickers guns to Arbil. The 2nd Battalion left Mosul on March 29th and the 3rd on the 30th. Column Commander and Staff joined at Arbil by air.

March 30th. April 5th. April 6th. April 8th. April 10th. April 11th. April 12th. April 15th. April 16th. April 17th.

The whole column, strength : Officers 25, Other Ranks 2,482 and animals 378, assembled at Arbil on April 5th. The weather was awful, and the country became water-logged. Intention was to march on April 7th, but the start had to be postponed first until the 8th and then until April 10th. On April 10th Frontiercol left Arbil and reached the Bustura Chai. Marched to Dera April 11th, and Duwin Qala April 12th. Here they were stuck in the heavy rain until 15th April, the camp being sniped at night. The column marched on the 15th and crossed the Hurash Chai, which was in flood, and the animals got bogged. They reached the Sorek Chai on the 16th and found the Kurds holding the Spilik Dagh. The Air Officer Commanding directed Frontiercol to remain facing the Spilik Dagh until Koicol could advance sufficiently to outflank them. For the present therefore Frontiercol remained in position. The 2nd Battalion on the left of Frontiercol encountered Kurds in Kani Chirgan and a good deal of firing occurred, but no casualties to the Levies. On the 17th April Captain Littledale, who was in the Dasht-i-Harir with his Police, encountered the enemy in Batas and had a few casualties.

April 17 and 18th. April 19th. April 20th. April 21st.

The R.A.F. bombed the Spilik Dagh on the 17th and 18th April. Koicol had an action on April 19th and were attacking Baijan Pass on 20th April. On the same day Frontiercol occupied the Spilik Dagh and marched to Qarachin. On April 21st they were at Kani Utman. Orders were issued for the attack on Rowanduz next day, but Euz Demir and the Turks left on the night of April 20th.

Rowanduz occupied. April 22nd. April 24th.

For the advance Lieut.-Colonel Barke took the 3rd Battalion, and linked up with Koicol, a company of the 4th Battalion climbed the Kurik Dagh and Rowanduz was taken on April 22nd. Reconnoitring detachments were pushed out and on April 24th Rab-Tremma Daniel Ismail had a small scrap, and killed one Kurd and brought in some cattle and donkeys.

Colonel Dobbin with Frontiercol remained in command at Rowanduz. General Vincent marched with Koicol to deal with the Sulaimani situation. A camp site was chosen near Rowanduz and a landing ground chosen and made at Diana, about three miles away. The 4th Battalion occupied a camp at Kani Utman.

The rest of the year was given up to reorganization and administrative matters.

Moves during 1923.

The 2nd Cavalry Regiment had now returned to Mosul. Two squadrons had taken over Zakho from the 2nd Battalion when it had marched off to take part in the Rowanduz operations. On the arrival of a battalion of the Iraq Army at Zakho, the 2nd Cavalry Regiment handed over to them and moved to Feishkhabur on April 1st. Here they remained until mid-June, and returned to Tank Hill Camp, Mosul. They had a very bad time from malaria.

The Remount Depot had been formed at Mosul and came under orders of the O.C. 2nd Cavalry Regiment.

The 1st Marsh Arab Battalion moved from Nasiriyeh to Baghdad, where they took over the Residency Guard and Air Headquarters Guard, and also provided one company on duty at rail-head, Shergat. 1924.

The Levy details camp, to which all personnel arriving for Levies had hitherto reported, closed down on February 29th. February 29th.

The 2nd and 3rd Cavalry Regiments were amalgamated during April and May and became the 2nd Cavalry Regiment, and moved from Mosul to Arbil. April and May.

In May also the Levy Pay Office moved from Baghdad and opened at Mosul on May 26th. May 26th.

A further reorganization was made in June, when the 2nd Line Transport of the four Infantry Battalions was pooled, and made into a Mule Transport Company under O.C. Remounts. This was found successful and finally adopted in October.

In July was started the scheme of enlisting Yezidis for the Levies. It was proposed to form a Yezidi Squadron, which was to form part eventually of the 1st Cavalry Regiment. The officers, who were to be specially enlisted, were to be either Assyrians or Yezidis, but not Moslems, for religious reasons. This scheme did not work, as the Yezidis proved far too difficult to train, and not very amenable to discipline; but they were good with animals. A number of them went finally to the 1st Line Transport of the 4th Battalion, and here gave some trouble, when out on the march. They were all discharged when the 4th Battalion was disbanded. July 30th. Yezidis enlisted.

It was found after some experience, that as a rule the Chaldeans did not make good soldiers, and they were gradually discharged.

The operations of this year will now be described. The 1st Cavalry Regiment began early in February, when a column went from Kirkuk into the area when the Jaf were camped to enable a peaceful settlement of taxes. This was successfully carried out.

CHAPTER VI

1923–1924

Sheikh Mahmud's Second Rebellion begins. 1923. May. 1924. May. Sulaimani bombed.

THIS year Sheikh Mahmud's actions in Sulaimani had brought matters to a head.

In May 1923, owing to his correspondence with the Turks, he had been ejected from Sulaimani by General Vincent, who had occupied the place; but after the British troops had left, he came back, in July 1923. He remained there throughout the winter, but by May 1924 his tyranny had become so excessive that the High Commissioner sent him an ultimatum, and on an unsatisfactory answer being received Sulaimani was bombed.

In May occurred the very unfortunate affair, called The Kirkuk Disturbance, the memory of which has not yet died down.

The Kirkuk Disturbance.

It must be remembered that the Assyrians are not popular in Iraq. Although some of the tribes had lived for centuries in the Sapna area, and others, such as the men of the Baz and Jilu tribes, often came into Mosul and other towns for employment, as a general rule the mountaineers of Hakkiari were complete strangers. Then they came in, in a body in 1918 as refugees, and created a difficult problem. While they themselves as mountaineers, disliked, and looked down on the town and plain dwellers of Iraq, these in their turn disliked, and to a certain extent feared the Assyrians. It must be admitted also that the Assyrians did not try and make themselves popular, not knowing how to do so. The natural antipathy accentuated by different religions continued and it only required something to start trouble.

This began in Mosul. In August 1923 trouble occurred in the meat market, which spread, and one or two Assyrian children were killed. No one was brought to book for this, and the Assyrians much resented it, and talked of the Iraq Government not administering justice in their case against Moslems.

In May 1924 the 2nd Battalion Iraq Levies was in Kirkuk, and their families with them. The Battalion was in the process of forming a camp at Chemchemal for the Sulaimani operations, and on May 4th only two companies and the Assyrian families remained in Kirkuk.

Already there had been a certain amount of back-chat between the townspeople and the Assyrians, in which the former, seeing the greater part of the Battalion moving out, threatened to deal with the Assyrian women when they had gone. Matters were in fact very tense. Many people in the town were in sympathy with Sheikh Mahmud.

At 09.30 on May 4th there was a disturbance in the bazaar. An Assyrian soldier returned wounded, after a dispute over the price of an article in a shop. May 4th.

Rab Khamshi Baijo went with the Regimental Police to clear the Assyrians from the bazaar.

The remnant of the Battalion was ordered on parade just as they were. Captain Growdon, the Police officer, arrived at the Levy Camp, and he and Captain P. P. King, commanding one of the two companies left in Kirkuk, both went on to the parade.

Just as they arrived, Rab Khamshi Baijo and the Regimental Police returned, bringing with them two more wounded men, and reporting that the bazaar was clear. The two wounded men said they had been knocked down from behind with heavy, or loaded, sticks, during a dispute. They also said people were calling out to them in the bazaar, "Now that half of you have gone to Chemchemal, we are not frightened of you."

Captains King and Growdon explained to the men on parade that there had been trouble in the bazaar. That it was out of bounds for the rest of the day. That the shopkeepers, who caused the trouble, would be arrested and tried. That the Battalion was on its best behaviour not to cause trouble. A Police piquet was to be placed on the bridge. Other men then brought forward complaints. Growdon and King began to go into the complaints, and the parade was dismissed.

On leaving the parade ground, the men had to pass by a Chai-Khana. Suddenly in answer to some remark by the people inside, a riot broke out.

The men rushed the place, and broke chairs over the heads of the people in the Chai-Khana. Then a small body armed with sticks made for the bridge. Captain King with the Native Officers, Baijo and Gewergis, rushed towards the bridge to try and stop the men, and Sergeant Burgess of the Police made for it in an arabana.

The Police piquet and the officers attempted to force the men back from the bridge, but some got over and were fired at from the other side. This scattered them and caused several casualties. The piquet of Assyrians on the west side of the bridge came under fire also. A number of the men ran back to their lines, and returned with rifles and S.A.A. and firing began in all directions.

Captain King caught as many men as he could, took them into Kirkuk

Fort, where their arms were taken from them, and they were put under guard, in the Cavalry Magazine.

Meantime one party had made their way across the bridge and captured a prominent building known as the House of Tooma, and took up their position on the roof. They could be seen here from the Kirkuk Fort, in which was Captain A. T. Miller, the Administrative Inspector, Kirkuk, with Captains King and Growdon.

They wired off to the Colonel-Commandant, who was staying with the A.O.C. in Baghdad. He came over by aeroplane, the armoured cars were sent for, a message was dispatched to Lieut.-Colonel G. C. M. Sorel-Cameron at Chemchemal and a wire of what was going on to Baghdad.

Captain O. M. Fry of the Levies was on the aerodrome, and received a message to come up at once. He made his way under fire from the houses of Kirkuk, near the Police Station, over the bridge to the House of Tooma, and after some time collected various parties of Assyrians. He eventually got together some eighty of the Assyrians, including three officers, and got them back to the fort, being fired at from the houses near the Police Station. Firing continued in the town. The Levies had lost five killed and seven wounded, and one civilian Arab employee killed. The casualties of the Kirkuk people were about fifty killed. About one hundred Christian refugees had assembled in the fort, and were kept there and rationed. The town was quiet by about 5 p.m., and at this hour a platoon of the Royal Inniskilling Fusiliers arrived by aeroplane, and took over guards and patrolling.

As the situation was most critical, and the feeling very bitter, the whole of the Assyrian Battalion was marched out of the town at once, families included, and camped at a place four miles out. Here they remained for a day, under Captain Fry's command, and then after collecting such transport as could be found, in the way of arabanas and animals, he moved a short way towards Chemchemal. Colonel Cameron left Kirkuk on the 6th May, took over the column and began the march to Chemchemal.

May 4th.

May 6th.

The column consisted of 23 old men, 404 women, and 172 children, escorted by two companies. They reached Chaman Bichuk on the 6th, and Qara Anjir on May 7th. Here at 6 p.m. the Kurds attacked one of the piquets, but were driven off, and one was seen carried away.

May 7th.

May 8th. Attack on Cameron's Column on the Kirkuk road.

Next day the column started at 08.30, and the Kurds attacked on all sides and kept on firing at the column until three miles from Chemchemal. The column lost one man killed and one missing. The Kurds were fired on by aeroplanes, and at one point the Assyrians succeeded in getting to close quarters, and inflicted a number of casualties on them, fifteen killed were taken to Kirkuk.

The Battalion remained at Chemchemal for the present.

The result of this affair was that a court-of-enquiry was held. After this three officers and twelve men were arrested, but after the trial the three officers and two of the men were found not guilty, and nine, who were found guilty, were put into Baghdad Jail. Results of the Kirkuk disturbance.

The killing of all these civilians could not of course be let pass without severe penalty, but it was realized that the Assyrians had great cause for irritation, and being fired into caused these very hot-headed people to act as they did.

Sir Henry Dobbs, the High Commissioner, promised the case of the men found guilty should be reviewed in nine months' time. After the good work by the 2nd Battalion in the operations against Sheikh Mahmud in 1925, a petition for their release was put up; it was put up again on 16th January 1926 on the conclusion of the frontier negotiations with Turkey, and again on the 15th April 1926, the Prime Minister, Sir Abdul Mushin Beg, being approached personally. The Kirkuk Prisoners.

The Iraq Government agreed to their release, and they arrived in Mosul on July 5th and went straight to Mai, where they were to live by orders of the Iraq Government. Eventually this restriction was withdrawn, and they returned to their own homes.

Captain Fry received the Military Cross for his gallant action during the critical period of the outbreak.

Although the incident is now long past and although it was ended by the Iraq Government releasing the prisoners, and by David D'Mar Shimun writing to thank the King, the High Commissioner, and the Prime Minister for their clemency, the outbreak is still remembered in Kirkuk, and small incidents from time to time show there is still a possibility of trouble.

In July after the bombing of Sulaimani it was decided that the place must be occupied, and the district brought into order, and that the Iraq Army assisted by the Levies, Royal Air Force, and armoured cars should carry out the operation. Operations for the occupation of Sulaimani.

On July 15th the 2nd Battalion (less two companies and Machine-Gun Section) marched to a camp three miles south-west of the Bazian Pass. The armoured cars arrived in Chemchemal. On the 16th the Pass was occupied and piquetted, the armoured cars went on a short distance, covering repair of the road, which was done by a working party of Captain McKay Lewis's "A" Company. This was to make the road passable for wheels. July 15th.

The Battalion camped at Kani Shaitan Hassan. The Iraq Army Column under Colonel Ali Ridha joined Colonel Cameron and the Levy Column at Kani Shaitan Hassan on July 17th. Air reconnaissances were made over Tasluja Pass. July 16th. July 17th.

July 18th. On July 18th, the march was continued on Tasluja Pass. The armoured cars went on, with a Levy working party for the road. On reaching the pass the Levies marched up both sides and piquetted them. The Iraq Army Column marched through, and camped at a point three miles along the Sulaimani road. They were sniped at night, and heavy rifle and machine-gun fire took place.

July 19th. Iraq Army occupies Sulaimani. Next day the Iraq Army and armoured cars marched on to Sulaimani and occupied it. Sheikh Mahmud retired to Barzinjah, and the 2nd Battalion Levies returned to Chemchemal.

During the operations the 2nd Levy Cavalry Regiment provided escorts for supply columns which reached the Iraq Army.

From the occupation of Sulaimani began the operations against Sheikh Mahmud which lasted for the next three years.

In August Lieut.-Colonel Cameron took a column from Chemchemal to deal with the villages of Trammal Uliya and Bagh, which had both been involved in Sheikh Mahmud's rebellion, and the latter was a favourite headquarters of Kerim Futteh Beg.

The Column consisted of :—

Two Squadrons Cavalry, Iraq Army.
H.Q. and Two Companies 2nd Bn., Iraq Levies.
One Section Machine-Guns, Iraq Levies.

August 21st. August 23rd. This column left Tasluja on August 21st, and burned the two villages on August 23rd. The only opposition was from small bodies of snipers, who fired on the Iraq Army Cavalry patrols, who attacked them, and killed one and captured another, and brought in two rifles.

August 25th. This column was back in Chemchemal by August 25th.

September. One company, 2nd Battalion, remained in occupation of Tasluja Pass, in the place of a company of the Iraq Army which was required for operations near Sulaimani.

September 22nd. It returned to Chemchemal about September 22nd. Colonel Cameron had already sited the defences of this pass, which were from now on held by one company, Iraq Army.

While these operations were going on, the High Commissioner asked for a demonstration march to be carried out towards the Persian Border about Pushtashan. This was carried out by the companies of the 4th Battalion from Kani Utman.

The point of interest now shifts to the North. For some time the Assyrians, particularly of the Tkhuma and Upper Tiari clans, had been very gradually percolating back into their own country over the Turkish Border.

This border was still undefined, and two Levy officers had already visited the Upper Tiari and Tkhuma people in their old homes.

Turkish attack on Hakkiari.

Some time in August the Turkish Wali of Julamerk made a tour into Hakkiari and came into collision with the Assyrians. Some firing ensued and the Wali's baggage was captured.

This drew the attention of the Turkish Government that way, and they resolved to take action.

September 13th. Turks invade Iraq Territory.

On September 13th, a Turkish force suddenly crossed the Hazil River and appeared to be threatening Zakho. Air action was taken against them next day.

September 14th.

September 15th.

On September 15th, the Turks attacked Bersivi, only nine miles north-east of Zakho. The Air Vice-Marshal placed Colonel-Commandant Dobbin in command of Mosul, Zakho and Amadiya.

At Amadiya was: One Company, 3rd Battalion Levies.
One Company, 4th Battalion Levies.
One Section of Machine-Guns.

Lieut.-Colonel Barke, O.C. 3rd Battalion, happened to be there inspecting and took command. There were rumours of a Turkish advance, and Mar Yoyallah, Bishop of Doura, reported firing going on near Ashita, and artillery in action. Barke sent two platoons under Lieut. Hart of the 4th Battalion to Ain D'Nuni.

Operations in Ain D'Nuni area.

September 17th.

On September 17th, Barke moved forward to Ain D'Nuni with the rest of his force, less two platoons which he left at Amadiya.

The same day, Turks crossed the Khabur near Merga, and an air patrol of No. 55 Squadron, operating from Mosul, was fired on, east of Chellek.

September 18th.

So far numbers of Irregular Assyrians had been at Ain D'Nuni, and available to hold up the Turks. These people's first concern, however, was the safety of their families, and they left Ain D'Nuni and went back with them to Bebadi. This left Colonel Barke and his small force isolated, with Turks in front and on his left flank. Orders were sent for him not to go more than one day's march from Ain D'Nuni. One of the most doubtful quantities of the area was Hajji Rashid Beg of the Berwari Bala Kurds. He had fought the Government in 1919 and 1920, and was believed to be pro-Turkish. Colonel Barke moved to Benawi on September 18th and sent Hajji Rashid Beg a message to come and see him. Hajji Rashid Beg replied he was sick and could not do so, an excuse which was probably untrue, and as matters were, was very suspicious. A party of seventy Assyrian Irregulars under Rais Iskaria joined Barke on this day.

September 19th.

On September 19th a wire was received from Constantinople to say that the Turkish Police were taking action against brigands, for the attack on the Wali of Julamerk. Later proofs were obtained to show a good deal more than Police were used. Barke moved his column to Ain D'Nuni, leaving Iskaria and his seventy men holding Benawi.

No. 55 Squadron bombed the Turks at Hauris, but the Turks took Ashita, and refugees came in to Ain D'Nuni on receipt of a wire of the situation from Barke. Two further companies of the 3rd Battalion were ordered by Levy Headquarters to march from Diana to Amadiya, and a

September 20th.

company of the 4th Battalion took their place on the 20th September. Two platoons, 3rd Battalion, under Captain Moody, were ordered to march from Mosul to Dohuk. The occupation of Ashita caused a general Assyrian

Retreat of Assyrians from Hakkiari.

retreat from their country, and a number of Maliks and others were now in Ain D'Nuni. Barke held a meeting with the Maliks and persuaded them to push out piquets towards Ashita, supported by two platoons of Levies under Lieut. Hart. Rais Iskaria and his seventy Irregulars still held Benawi on Colonel Barke's left. Parties still held Jebel Zawita, Desht, Aden, and Gali Sarhara. Malik Khoshaba still remained in Lizan. Shamasha Yonan was reported surrounded by Turks on the Walto Dagh, and a hundred men with him. Aircraft passing over the area were met by artillery and rifle fire; they did good work bombing and enabled the

September 21st.

Assyrians to occupy Zawitha village. At 01.30 hours on September 21st Barke was awakened by Mar Yoallaha, the Bishop of Berwar, to tell him that the Irregulars had gone with the exception of Rais Iskaria and his seventy men at Benawi. His orders were to hold on, but not to get cut off, and in view of that order, and the report that Hajji Rashid Beg was definitely hostile, he withdrew to Bebadi at 14.30, after sending orders to Iskaria to leave Benawi at 13.00, and to O.C. Detachment at Bebadi to send a platoon to piquet the top of Ser Amadiya.

Two platoons of the 3rd Battalion were ordered by Levy Headquarters to Dohuk.

September 22nd.

On September 22nd, the Levy Pack Battery was ordered to start for Amadiya, escorted by a troop of the 1st Levy Cavalry. Colonel Barke was ordered to continue to hold the Ser Amadiya, and the Assyrian Maliks were ordered to join him with as many men as possible. Two platoons of the 4th Battalion were sent to Aradin as a flank protection, and at Aradin Islam, the upper or the Kurdish part of the village, they had some slight opposition. Captain Moody and the two platoons 3rd Battalion from Mosul were ordered from Dohuk to Amadiya. A reconnaissance to Berwar

September 23rd.

on September 23rd found villages in that area burnt. The Levy Column of two companies from Diana, under Captain J. O. Watt, reached Rezan

September 24th.

this day. Next day Flight Lieutenant Reed, who was this time acting as Intelligence Officer, with Lady Surma, got four hundred men up to join

Barke. Lady Surma was the moving spirit and harangued the Maliks and other Assyrians, urging them to go forward to action against the Turks. She was mentioned for her work and was awarded the M.B.E. A small post consisting of Levies, Police and Irregulars was established at Hayas. A patrol from this post had an encounter with the Turks, and two Police missing.

Lady Surma urges the Assyrians to meet the Turks.

On September 25th Colonel Barke advanced from Ser Amadiya. The Irregulars were formed into three columns, and stiffened by two platoons of the 3rd Battalion, all being under Lieut. R. Hart and R.S.M. MacGregor. The Bishop Mar Yoallaha, handing his cassock to his deacon, went in with the Irregulars. This force advanced on Ain D'Nuni, lighting fires as they advanced to show their progress. They went through Hayas, where they encountered the enemy, and drove them back, and by night-fall they were in occupation of Benawi, Mai and Ain D'Nuni, with a loss of one killed and two wounded. Barke established his Headquarters at Ain D'Nuni. The two thousand refugees in the Sapna, who had no food, caused some trouble by looting villages about Suara Tuka, and had to be rationed. The 26th was quiet. On the 27th September, Captain Moody with two platoons from Mosul arrived, and also the column of two companies from Diana, under Captains Watt and J. Moulding. They immediately took over the piquet line from the Irregulars, who were now placed in support.

September 25th. Advance from the Ser Amadiya.

Action near Hayas.

September 26th.

September 27th.

The Levies and Irregulars continued to hold this line. Refugees still continued to come in. The Malik of Baz and sixty men arrived from Dohuk on October 1st, and three men and one woman got in from Darawa on the Walto Dagh on October 4th. No. 55 Squadron kept up close reconnaissance of the Turks and were invariably fired on, and on October 3rd and 4th bombed the Turks on Ashita. Barke's force had small opposition at Beidu on October 2nd and 3rd. Turkish prisoners taken were handed over to the Qaimakhan of Amadiya, and also Hajji Rashid Beg, who had been arrested for his part in the affair. The Turks occupied Chal on October 5th. On October 8th Captain Baddiley noticed that the hill of Zawitha was now unoccupied, and went forward with two platoons and occupied it. This proved waterless, and he withdrew to the old line, having one man wounded. The same day the Turks drove the Tiari men from Naramik, and occupied it.

October 1st.

October 4th.

October 3rd and 4th.

October 5th.

October 8th.

October 9th.

This ended the affair. On October 11th Captain Baddiley rode with a letter to the Turkish Officer at Ora from the High Commissioner, through the Turkish lines. He was allowed to proceed as far as Ora, where he delivered his letter to a junior officer, who said he had orders to let him come no further. From now on fighting ceased, except an affray between the Police and Turks at Chellek, where the Turks had one killed, and eight

October 11th.

October 30th. captured. Of the Assyrians who had been in Turkey when this affair began, Shamasha Yonan died of his wounds received at the Walto Dagh on October 25th. He had put up a very stout fight against the Turks. In anticipation of trouble he had rationed a cave on the Walto Dagh, and when the rest of the Upper Tiari left he held on with some seventy men, and a few women and children. Unfortunately the water was outside the cave, and the Turks trained guns and machine-guns on to it. He refused to surrender though called on. When matters became desperate they decided to leave and get out of the cave by night. Meantime the Irregulars with Barke asked for Levies to go with them and rescue Shamasha Yonan. Barke replied that Levies could not invade Turkey in uniform. A party of Irregulars therefore crossed the border, got in touch with Shamasha Yonan and brought him in with the rest of his party. Four men and twenty-four women were captured in the caves, and sent in to Julamerk.

October 25th. Shamasha Yonan's defence of the Walto Dagh.

November 21st. The Levy Pack Battery and Headquarters of the Machine-Guns returned to Mosul, leaving on November 21st, and on the same date the company of the 4th Battalion left for Kani Utman. The two companies 3rd Battalion and section of the machine-guns went to Bebadi, and these with a wireless detachment with R.A.F. personnel formed the garrison at Bebadi for the winter. Two platoons under R.K. Yokhannis Nuriya remained at Ain D'Nuni for the winter.

This affair was probably the surest test of discipline for the Assyrian Levies that they could have had. They could see their homes going up in flames and they did not know for a certainty if their relatives and families were safe. But they carried out their orders under the British Officers without question. This point has been fully brought out in the High Commissioner's report of the operations.

The following received special mention for their action in this operation. Captain J. O. Watt, Lieuts. R. Hart, P. J. T. Baddiley and J. R. Bourne, Sergeant-Major H. J. Edwards, D.C.M., and Sergeant E. H. Riches. Rab-Emma Shain Gewergis and Rab-Khamshi Zia Gewergis, and Ibrahim Effendi, Commandant of the Police at Ain D'Nuni, and, as has been mentioned before, Lady Surma D'Beit Mar Shimun.

During 1924, three Officers, Captains W. McWhinnie, M. Hammond, O.B.E., and H. E. Bois, and also Sergeant Dawson died of disease contracted in the country.

CHAPTER VII

1925

THE year 1925 turned out to be a very active one for the Levies. The situation of the units at the start of the year was as follows :— 1925.

1st Levy Cavalry Regiment—Arbil.
2nd ,, ,, ,, —Kirkuk.
1st Marsh Arab Battalion—Baghdad, with one Company at Shergat.
2nd Battalion Assyrians—Chemchemal—one Company Koi Sanjak.
3rd ,, ,, —Aqra, with two Companies Bebadi.
4th ,, ,, —Kani Utman—one Company Rowanduz.

Levy Headquarters, The Pack Battery, Levy Pack Ambulance, Depot, Transport and Remounts were all at Mosul. Headquarters Machine-Gun Company at Mosul with Sections at Bebadi, Chemchemal, and Kani Utman.

The winter of 1924–25 was a hard one in Iraq, and the cold killed many of the overworked and under-fed transport animals. The crops were late in coming up owing to the long winter, and the locusts, which were bad that year, caught the crops as they were coming up. The result was many people were ruined, and the price of cereals became very high. It had the effect of many people taking to highway robbery and joining the rebellious bands of Sheikh Mahmud and his adherents. Effect of hard Winter and Locust plague.

Sulaimani was the chief centre of interest. Fighting began there early, in fact it hardly ceased throughout the winter, and it went on the whole year. Sulaimani Fighting begins.

The Iraq Army had occupied Sulaimani in July 1924. They did a certain amount of patrolling in the area, but they had two serious set-backs, one at Kurmal and one close to Qarachugan, which encouraged the Sheikh Mahmud's party. He began to close round the town, and by March it was to all intents and purposes closely invested, as no one except with a strong escort could go outside the place. Moreover, Sheikh Mahmud's adherents were active inside the place, and the Mutasarif Ahmed Beg-

Taufiq-Beg, himself a Kurd of the Pizhdar tribe, had some twenty shots fired into his house one night, luckily without damage. The Special Service Officer, Flight Lieutenant A. McGregor, D.F.C., was nearly shot one night in his house, by a man in the road below. Finally, Mr. Chapman, the Administrative Inspector, represented that a stronger body of troops was required to keep the area in order.

The Levies march to Sulaimani.

On April 22nd Lieut.-Colonel Cameron marched to Sulaimani with two companies of the 2nd Assyrian Battalion, leaving one company at Koi Sanjak, and Headquarters and one company to look after the camp at Chemchemal.

Yezidi operations.

April 18th.

North of Mosul operations had already taken place. The first action was against the Yezidis of Sinjar. This was carried out by the R.A.F. and armoured cars, and unfortunately Flight Lieutenant R. W. Pontifex and his mechanic were shot down and killed at Merkan.

Doski Kurd operation.

The Doski Kurds in the hilly area north of Simel then defied the Government. There was already one squadron of the 1st Levy Cavalry Regiment holding Dohuk, and another had just come up to relieve it.

Lieut.-Colonel Browne was in command of the Levies, in the absence on leave of Colonel Dobbin, and he took command of the operation, using the two squadrons of the Levy Cavalry to move from Dohuk and block the exits eastwards, one company 3rd Battalion under Lieut. Hart advanced northwards from Simel, and one company 4th Battalion with Lieut.-Colonel Gillatt and Lieut. Wallace came along the Sapna, and attacked from the north. A party of Police under Captain Sargon also took part in the operations.

The whole of No. 6 Squadron and one flight of No. 30 Squadron were placed at the disposal of Colonel Browne. However, all the villages except one surrendered, and that one, Hojawa by name, was bombed. The troops advanced, met no opposition, completed the destruction of Hojawa, and the operation ended with the surrender of most of the Kurds, while the chief one who caused the trouble fled to Turkey.

The point of interest now shifts to Sulaimani. On Colonel Cameron's arrival in that place, the officer in command of the Iraq Army troops was Ali Ridha Beg el Askari, brother of Jaafar Pasha el Askari, and ranking as a full colonel. But the situation, where such a case arose as was now in Sulaimani, was dealt with under Article IX of the Treaty of Versailles, which ordains that in operations where H.M. Forces take part with Iraq forces, the command shall be vested in the British Military Commander subject to any special arrangements. This was confirmed by a War Office Order which laid down that under no circumstances will British Officers, or troops, be put under an Arab Officer.

May. Cameron in Command at Sulaimani.

Therefore on getting to Sulaimani, Lieut.-Colonel Cameron took command of all troops in that place, acting directly under the orders of the Air Vice-Marshal, who directed the main object of the operations from Baghdad. On May 2nd a conference was held at Kirkuk under the High Commissioner and the action to be taken for restoring order in Sulaimani area was decided on. The first operation by Lieut.-Colonel Cameron was to clear the Kurds from the immediate vicinity of Sulaimani and to build the defences round the place. The Azmir Dagh was cleared of Kurds by three columns of Levies and Iraq Army, and a post was sited and constructed to keep the Azmir Pass open. One company of the Iraq Army with Captain J. M. Coffey, M.B.E., Royal Inniskilling Fusiliers, Liaison Officer with the Iraq Army, remained up there to complete the work.

Azmir Dagh cleared.

May 15th. 2nd Levy Cavalry leave for Halebja.

The Air Vice-Marshal decided that Halebja must be occupied to complete the settlement of the area, and on May 15th Lieut.-Colonel B. T. T. Lawrence, V.C., O.C. 2nd Levy Cavalry Regiment, marched with that object from Kirkuk. A company of the 2nd Bn. Bedfordshire and Hertfordshire Regiment took his place.

May 17th.

On May 17th Lieut.-Colonel Lawrence, then on the march beyond Chemchemal, heard of rebel Kurds at Kanisri, and detached one squadron to try and catch them, but the rebels made off.

May 18th. Dar-i-Qeli Operation.

On May 18th Colonel Cameron, acting on information that Kerim Futteh Beg, the Hamawand rebel, was in Dar-i-Qeli, attempted to surprise him by night.

May 18–19th

He marched himself with a mixed Levy and Iraq Army Column, sending an Iraq Cavalry detachment with Captain J. H. Gradidge, O.B.E., Liaison Officer with the Iraq Army, and asking Lieut.-Colonel Lawrence to co-operate with one squadron.

This operation was within very little of being successful, the Kurds were seen making off before the two bodies of Cavalry, who did not know the ground, had reached their position. Darkness, unknown country and the doubtful loyalty of a guide, caused the failure to capture the leader, but the Kurds had two killed, one wounded and eight prisoners taken in an action at Dar-i-Qeli.

May 19th. May 20th.

Cavalry attacked at Arbat.

The 2nd Cavalry Regiment reached Sulaimani on the 18th May, and Arbat on the 20th. Lieut.-Colonel Lawrence camped at a place some 1,200 yards north-east of the village. So far no Kurds had been seen; but, as soon as it was dark, they got into Arbat and on to the mound near it, and began sniping the camp. Some men and animals were hit at once, and Lieut.-Colonel Lawrence directed Captain S. Fosdick to clear the mound with his squadron. He did this by a dismounted attack, and captured nineteen prisoners. The Levies had one killed, three wounded, and five animals hit. There was intermittent fire during the night; but in the morning no sign of the enemy.

May 21st.

Next day the march was resumed to Sarao. At Mawan, or Mohun, where there is an ancient fort of the mound type, said to have been made by the ancient Assyrians, the road turns north-east, and after passing between low hills and the village of Giryazah, is in a small open valley about four miles long and two broad. It is more or less oval in shape. The hills bounding it are low, and run gently down into the valley, except on the south side, where there is a rocky pointed hill some 300 feet above the valley. The entrance and exit of the valley are both narrow. At the Sarao end in particular, the road runs through a small pass barely 100 yards across. A shallow wady begins near Giryazah and runs diagonally across the valley with many twists and turns and leaves the valley with the road at the Sarao end.

Action at Giryazah.

The Regiment entered the valley and passed through Giryazah. So far a few Kurds had been seen on the rocky pointed hill on their right and south flank.

Aeroplanes of No. 1 Squadron (Snipes) came over, but the Kurds probably retired into the caves on the hill. At any rate only a few more men were seen, and the column continued the march.

Orders had just been given by Lieut.-Colonel Lawrence for one squadron to move on and seize the low hills on the right of the line of march and another to occupy the hills in front, which ensured the exits from the valley, and the rear-guard commanded by Lieut. W. Fuller Brown was just about Giryazah village, when the Kurds appearing from the direction of Mawan suddenly poured a very heavy fire into the rear-guard. At the same time they appeared in great numbers on the hills both on the right and left of the column.

The column had with it not only A.T. Carts but also a good deal of civilian pack transport with supplies chiefly belonging to the civil contractor. In addition to these, judging by the claims put in by the contractor, there was a great deal of extra stuff, very possibly stores for starting trade in Halebja. These stores were mainly on donkeys.

Directly the firing broke out the mass of transport animals and their drivers were in confusion. First of all a large number tried to get back towards Giryazah. In doing so they got mixed up with the rear-guard. Then, finding the Kurds coming on, the whole crowd came down in among the column. In the confusion that ensued, the Kurdish prisoners captured at Arbat escaped from the small dismounted guard escorting them, the guard being either killed or wounded.

May 21st.

Lieut. W. Fuller Brown, 4th Hussars, with his rear-guard, held off the attacking Kurds, and two mounted troops got among the disorganized transport and kept it moving on into the wady which ran down the valley.

In front the squadron under Captain R. C. Hill, M.B.E., sent by Lieut.-Colonel Lawrence to seize the low hills at the exit, found it held, and was driven back by the Kurdish fire.

Colonel Lawrence then ordered the whole force to take cover in the wady, where he personally restored order among the panic-stricken civilian transport, and reorganized his troops for future action.

The wady gave some cover, but was partly enfiladed, and the force was still suffering casualties.

Colonel Lawrence then sent Captain Hill to make a mounted attack from the wady and clear the hills enfilading it, from the right, and two troops under Zabit Nuriman Effendi to clear the hills on the left front, that is the southern side. Both attacks, carried out at a gallop, were successful, and the Kurds retreating before Captain Hill's attack were caught in the open, and machine-gunned by No. 1 Squadron from Sulaimani. This was followed by a mounted attack by Major Fosdick's Squadron on the hills on the right front, which the enemy vacated without resistance.

The march then continued, Lieut. Fuller Brown continuing to hold off the Kurds, who kept up fire at long range. The Regiment reached Qara Teppeh that night. The losses of the Levies were 9 killed, 1 missing and 18 wounded, while 30 animals were killed, 9 missing and 29 wounded. Great quantities of material and supplies were lost from captured or killed transport animals, and some A.T. Carts which were overturned in the wady, or had the animals drawing them killed.

They reached Halebja next day without opposition. Here they remained May 22nd. until September. In the action of May 21st the Sub-Assistant Surgeon Gul Akbar Shah and the Veterinary Surgeon Nur Mohammed were both awarded the M.B.E. for gallantry in action. The latter though wounded, and with broken ribs, continued his work among the animals during the fighting. In addition Zabit Abid Abdullah Effendi and Zabit Majid Mousa Effendi, though both wounded, displayed great gallantry in this action. The assistance rendered by Flight Lieutenant Luxmore and No. 1 Squadron was brought to the notice of the Air Vice-Marshal.

The force which attacked Colonel Lawrence consisted of Sheikh Mahmud's own adherents, some of the Hamawand under Saber, son of Kerim Futteh Beg, contingents of the Shatri, Roghzadi, and Haruni Jaf, and a number of local Kurds from villages round. Colonel Cameron marched from Cameron leaves for Halebja. Sulaimani at once, to deal with the area.

He left on May 22nd with a mixed column for Halebja, which he reached May 22nd. on May 24th, following the line of Lieut.-Colonel Lawrence's march. He May 24th.

spent the 25th and 26th camped at Halebja, and gathering information, and on the 27th May started back vìa Hayas, Mawan, Bezancur and Arbat, burning the villages of those implicated in the attack of Lawrence's Column and rounding up all stock. The only opposition was at Hayas, where a cavalry patrol of the Iraq Army was fired on, one horse being killed.

May 27th.

Punishment of rebellious villages

May 30th.

Cameron's Column reached Sulaimani on May 30th. The town was very disturbed, and shooting inside went on most nights. A complete comb-out was done by the troops and Police, and matters improved.

One new plan was adopted at this time which was this. To raise among the Assyrians a body of Irregulars 200 Mounted and 50 Dismounted. Their mission was to hunt down Sheikh Mahmud and his most prominent adherents.

Assyrian Irregulars raised.

These Irregulars were duly raised and armed, and sent off to join Cameron at Sulaimani. They did not for various reasons prove a success, the chief one being that they were strangers working in a country of people of a different nationality to themselves, and generally in sympathy with Sheikh Mahmud. At any rate a mixed mob of men, dressed in costumes varying from complete native Assyrian dress to a man in complete European dress with a grey bowler hat, paraded before the Colonel-Commandant at Mosul, and were dispatched to join Cameron at Sulaimani. Three Assyrian Officers, Rab-Khamshis Maxut Niko, Raoul Yokhanan, and Ozario Tamras, all of the 2nd Battalion, were chosen to command the Irregulars, and Lieut. Fairrie, Iraq Levies, in command of the whole force. His account of them is as follows: The Irregulars were only raised by increasing their pay. They had no idea of discipline on the march, or in the field. During the action at Kinaru on June 25th, they fought each man for himself, bravely enough, but with no regard for the rest of the force (as will be shown in the description of this action) when put on piquet near Penjvin, they all left the piquet line and descended into the village to loot. They did excellent work on any independent patrols and always accounted for Kurds. Finally when ordered to remain in occupation of Chaortah, they declined to do so, except under conditions which could not be carried out. This was in contravention of what they had agreed. Lieut. Fairrie, David D' Mar Shimun, and the Colonel-Commandant all interviewed them, and finally sixty-nine of the men agreed to obey Government orders. These were formed into a mounted body called locally "The 69th Light Horse" under Rab-Khamshis Maxut Niko and Raoul Yokhanan. The rest were sent back to Mosul and discharged. The 69th Light Horse were kept on until the end of the year, working under the Administrative Inspector, Captain W. A. Lyon. While Cameron had been away dealing with the Kurds, who had attacked Lawrence's Column, work had been continued on the Sulaimani defences. A post was also being constructed on the top of the Azmir Dagh, just off the road over the Azmir Pass, by a company of the Iraq Army. Captain J. M. Coffey, M.B.E., Royal Inniskilling Fusiliers, one of the British instructors of the Iraq Army, was with this party. They were continually sniped at night.

Azmir Dagh

On the night of June 4th a body of Kurds consisting of men of the migrant Ismail Uziri tribe, a branch of the Jaf, and locals from Sitak got close round the post and kept up a continuous fire. Captain Coffey went out on a personal reconnaissance, and was mortally wounded by a sniper, and died in a few minutes. The Iraq Army erected a small cairn to his memory on the spot where he fell. The Ismail Uziri went off to Persia and could not be got at, but Sitak was bombed by the Royal Air Force.

June 17th. Attack on Azmir Dagh Post. Captain Coffey killed.

On June 6th Cameron left Sulaimani for Chaortah, with the object of installing the Mudir there, and arranging for the future action of the Irregulars in the area, and dealing with any bodies of Sheikh Mahmud's forces who might be met.

June 6th. Cameron's Column marches to Chaortah.

He took with him the column as follows:—

2nd Battalion Iraq Levies (less 2 Companies).
One Section, 3rd Cavalry Regiment, Iraq Army.
2nd Battalion, Iraq Army (less 2 Companies).
W/T Detachment.
50 Assyrian Irregular Infantry.

Captain R. Merry commanded the Levies, and Lieut.-Colonel Hassan Hilmi the 2nd Battalion Iraq Army. No. 1 and No. 30 Squadrons co-operated.

He marched up the Azmir Dagh in three columns, to be in position to deal with any Kurds who might oppose the march, and concentrated at Sitak on the evening of June 6th. Here he was met by a friendly Pizhdar Agha Abbas-i-Selim Agha with 400 Cavalry and 200 Infantry. Cameron told him to place himself next day across Sheikh Mahmud's line of retreat to Persia by mid-day. This he did, but as the column were advancing on Chaortah, and before they deployed for attack on June 8th, Sheikh Mahmud, with 300 men, withdrew to Sirkan and the column occupied Chaortah.

June 8th.

A hostile Pizhdar lashkar was reported at Mawit. On the 12th June the column carried out a reconnaissance through Naomerak and Muberah, when the flank guard met Kurds near the former place and inflicted a loss of two killed.

June 12th.

On the 19th June Lieut. Fairrie joined the column with 100 mounted and 100 dismounted Assyrian Irregulars. Sheikh Mahmud was still reported about Chaman and Nurak with 600 men, and about to join the Pizhdar,

June 19th.

who were said to be 800 strong under Abbas-i-Mahmud Agha. This he did on the night of June 20th/21st. On that day and on the 22nd June reconnaissances were pushed out, the only contact with enemy being at Harman on June 20th. On the 23rd June, the Kurds sniped the Azmir Post on Cameron's line-of-communication.

June 20th.

June 23rd.

June 25th.

Action of Kinaru.

On June 25th Cameron moved out with the object of surprising the enemy outposts on the Siwel River, and ascertaining the enemy's strength. He took with him :—

> Headquarters and Battalion Levies (less 2 Companies),
> One Section 3rd Regiment, Iraq Army Cavalry Regiment,
> Headquarters and One Company 2nd Battalion Iraq Army Infantry,

and 250 Assyrian Irregulars, 100 mounted and 150 dismounted, under Lieut. Fairrie, with R.K. Maxut Niko, and Ozario Tamras under him. The entrenched camp at Chaortah was left guarded by one company Iraq Army and had sangars and trenches made for defence.

June 25th.

The column starting at midnight June 24th/25th reached the Siwel River at 04.30 June 25th. The leading of the column by the guide was in Cameron's opinion deliberately wrong, but as he had studied the map they reached the destination. The Irregulars under Lieut. Fairrie crossed, supported by "B" Company of the Levies.

The surprise was complete, and Kinaru was occupied and found deserted. Attempts to push on met with heavy fire and No. 30 Squadron aeroplanes were heavily fired on from the wooded hills. The Kurds pushed across the Siwel River on both sides of the column and seemed to be trying to cut off its retreat. Cameron's intention had been a surprise attack, and then to drop back on his trenches and sangars at Chaortah, and fight the enemy there if he came on. Retirement was ordered at 06.30 but the Irregulars who were well engaged with the Kurds, could not be got back until 09.00. During this period one machine of No. 1 Squadron was brought down and had to be burnt, but the pilot was rescued and the gun and instruments were saved.

The column returned to the camp by 12.15 and the Kurds pressed on after them.

Two sections of the Iraq Army Company left as camp guard made an attack with bombs on some advanced bodies of Kurds and drove them back with loss at 12.00, and later at 16.00 ; R.E. Ozario Tamras, 2nd Battalion, counter-attacked the enemy with his Irregulars, killing five. At 19.30 the enemy got into one of the sangars, but R.K. Sliman Sliwo re-took it with a bayonet charge, killing five Kurds inside. The fight ended at 23.30 hours. The column had lost three killed and twelve wounded.

June 25th.

There was no sign of the enemy next day, and the column rested. On the 27th, except for firing on the machines of No. 30 Squadron, nothing was seen of him. The Irregulars burnt Kinaru, and tried to draw the Kurds after them, while Cameron held a strong party concealed to await them. They would not come on. The report was that Sheikh Mahmud had gone back to Persia. June 26th. June 27th.

The column left Chaortah on July 5th, and marched via Chamak, arriving at Penjvin on July 6th. The advanced guard had a small action with Kurds. The town was empty. July 5th. Penjvin occupied. July 6th.

On July 8th the column left Penjvin, and marched to Serkan ; Kurdish snipers appeared on the flanks and lost one killed. As they hung round the camp Rab-Tremma Yakub Ismail took out a party of Levies, and caught them by surprise, killing two and wounding two more. July 8th.

Next day the column reached Barzinjah, accounting for two snipers on the way, and, on July 10th, got back to Sulaimani. The Kurds again sniped the flank guard on the Azmir Pass and lost one man. July 9th.

For these operations Colonel Cameron received a C.B.E. and mentioned for good work the following Levies :—

Captain R. Merry, Lieut. Fairrie, R.K. Ozario Tamras and Sliman Sliwo. He also mentioned Captain J. H. Gradidge of the Guides, attached to the Iraq Army, and Colonel Hassan Hilmi Beg, who commanded the 2nd Battalion, Iraq Army, and four Other Ranks.

On July 31st the company at Chemchemal was required to join the Battalion in Sulaimani. Two platoons of the company at Koi Sanjak therefore left that place and went to Chemchemal on July 31st, leaving one company looking after both Koi Sanjak and Chemchemal. The company from Chemchemal arrived at Sulaimani on August 9th.

CHAPTER VIII

1925

1925. AT the end of August, Colonel Commandant Dobbin returned from leave, and Lieut.-Colonel J. G. Browne took command of Sulaimani *vice* Lieut.-Colonel Cameron who went on leave.

The situation in Sulaimani, which had improved after Cameron's operations in May and June, suddenly became worse.

Sheikh Mahmud's emissaries, aided by Kerim Futteh Beg and his body of rebels, caused a rising in the area between the Qaradagh and Baranand ranges. Qaradagh village was attacked, the Police driven out, and the houses of Government supporters were burnt. Orders came from the Air Vice-Marshal for a column to march into the area to restore Government prestige there, instal a new mudir, and site a Police Post at Qaradagh, to overawe the settled Jaf in the Baranand area, and site Police Posts near Sarao and Sayid Ishaq.

The Column consisted of:—

2nd Battalion, Iraq Levies (less 1 Company).
4th Battalion, Iraq Army (less 2 Companies and 1 Platoon).
Detachment W/T. Section, Royal Air Force.

Column to the Qaradagh.

One flight, No. 1 Squadron, R.A.F., from Sulaimani, and one flight, No. 30 Squadron from Kirkuk, co-operated with the column. Six days'
August 20th. supplies were taken. The column left Sulaimani on August 20th, and went by the Baranand Pass, reaching Qaraman on August 20th and Qara-
August 21st. dagh on August 21st. Air action had been already taken against villages round. Only a few snipers appeared and these fired at long range at the tail of the column.

The column remained two days in the area and burned the houses of those implicated in the rising. No mudir was available, so the column continued with the rest of its objective. In going through Ribata Pass there
August 23rd. was some opposition and two Kurds were killed. On reaching the valley of the Tanjero River, a squadron of Iraq Army Cavalry with further supplies met the column.

On August 24th the column marched through the Jaf villages towards Halebja, being met near Zeleresh by a squadron of the 2nd Levy Cavalry Regiment from Halebja. The Kurds kept along the Baranand and sniped the right flank of the column at long range, but attempted no more.

August 26th.

The column completed its work and returned to Sulaimani on August 26th. In the absence of the column, a party of Kurds came over the Gwezha Pass and carried off some sheep and goats grazing near by. The Police tried to get them back, and had one man wounded. Raids of this sort were annoying, and any slight success encouraged the enemy, therefore leave was obtained from the Air Vice-Marshal to use raids also, when and where objectives could be found. The first of these raids was carried out on the night of August 27th/28th against the villages of Waldana and Werderlar, which had been the places from which the raid by Kurds of a few days before over the Gwezha Pass, had started.

August 27th. Raid on Waldana.

The raiding party was one company, 2nd Battalion Levies, and one company of the Iraq Army. The route lay over the Gwezha Pass. To hold the pass while the raiding party dealt with the villages, one squadron, Iraq Army Cavalry, went with the column.

As an additional precaution to prevent anyone getting wind of the raid and going off to warn any Kurds near, immediately it was dark, all available troops not on duty moved out and formed a circle of small posts just outside the town. They were to remain for four hours, by which time the raid could not be interrupted.

A very complicated evening and night ensued. The first occurrence was that some man lay in the road outside the Special Service Officer's house, with the intention of shooting him with a revolver as he came out.

A servant opened the door, however, and the man fired two shots through it, both of which missed, and fled.

The column assembled and marched through the piquet line at 22.00, *en route* for the Gwezha Pass. It seems that Sheikh Mahmud, or his supporters, had decided on this same night to shoot up Sulaimani, or to raid the place. At any rate, firing suddenly broke out in the piquet line against men coming on, who suddenly ran into it. This was kept on for some minutes, when the column on the move for the Gwezha was fired on in rear. This ceased, and the column continued its march, but it appears that the Kurds having located the column began to climb up the Azmir Dagh, on both sides of the column. Two parties of Levies were already climbing the same spurs, so the march continued, with the column in the centre, a party of Levies moving up to the top on ahead on each flank, and a party of Kurds on each flank of the column following the Levies. There was a searchlight in Sulaimani, and, owing to the firing, this was now turning its beam towards the Gwezha Pass. This showed the column up, which the Kurds then shot at. On reaching the top, as no further shooting occurred, it was decided to continue the operation. The Cavalry Squadron took up a position in sangars on both sides of the Gwezha Pass, and the column went on. The Kurds appeared again, making for the top of the Gwezha, were met by fire from the squadron,

August 28th.

and retired, and camped in a wady beyond the Azmir Dagh, where next morning they were located by the No. 1 Squadron reconnaissance and

August 30th.

well shot at. The column reached Waldana, burnt it and returned. Two days later the Iraq Cavalry made a sudden day raid on Werderlar.

September 6th.

On September 6th the Iraq Army Cavalry turned out suddenly in the afternoon and raided Kani Goma supported by No. 1 Squadron, on the information that Kerim Futteh Beg was there. Levies only acted in support. The report of the presence of enemy proved untrue.

September 7th/8th. Raid on Kani Darka.

On the night of September 7th/8th a raid was made on Asaban, Kani Darka, and Bazzanian. The column was two companies, 2nd Battalion Levies, and two companies of the Iraq Army. The route lay again over the Gwezha Pass, which was occupied until the column returned.

Asaban was surrounded, and two platoons of the Iraq Army left to hold it, and the column went on to Kani Darka. The Kurds were just moving into positions round the village, when the leading company of Levies under Captain K. F. McKay-Lewis, was making its way round the village to encircle it, and to join hands with the Iraq Army Company sent round the other side. A fight in the dark followed, in which the Assyrians got into the Kurds with the bayonet. A certain number succeeded in getting away through the gap between the Levies and Iraq Army

September 8th.

to Shukeh. Firing lasted half an hour. Twenty-two dead Kurds were picked up, and seventeen rifles brought in. No. 1 Squadron were over at dawn and, being shot at from Shukeh, bombed the village. Kani Darka was burnt. The column came back through Bazzanian, where some firing occurred, and two more enemy were killed. The Iraq Army platoon had been attacked by mounted Kurds, while the column was at Kani Darka, but had driven them off. One Assyrian Levy was killed. The enemy had 24 killed, 4 wounded were brought in, and 24 unwounded prisoners.

September 16th/17th. Raid on Bagh and Kupala.

The next operation was a raid on Bagh and Kupala, selected because it was known to be a place where Kerim Futteh Beg very often stayed. Both these places lie north of the Tasluja Pass and some eighteen miles from Sulaimani.

It was to be a combined operation for two columns:—

EAST COLUMN. Two Companies, 2nd Battalion, Levies.
One Squadron, 3rd Iraq Cavalry Regiment.
Two Companies, Iraq Army.

WEST COLUMN. Two Squadrons, 2nd Cavalry Regiment, Iraq Levies.
One Squadron, 3rd Iraq Cavalry Regiment.

East Column assembled at Tasluja on September 16th, and West Column at Kani Shaitan on the same date. September 16th/17th.

The objective of East Column was Bagh, and that of West Column was Kupala.

To draw attention from these objectives, the Administrative Inspector, Captain W. A. Lyon, had questions asked in Arbat concerning supplies in the area, and guides to meet a column moving on Qaradagh. East Column assembled on September 16th at Tasluja and marched at midnight 16th/17th on Bagh. They were just short of being successful. Kerim Futteh Beg left twenty minutes before the column was round the place. His rear-guard just escaped and made off through the rocks, leaving four ponies in the hands of the column. There was some firing. Bagh was burnt, and a village called Girga next to it also.

The Cavalry Column occupied Kupala, and also searched other villages round. There was no opposition, nine rifles were brought in.

Prior to this operation two platoons, one of Levies, and one of Iraq Army, were sent under command of Rab Emma Shimoel Tiya, 2nd Battalion, Iraq Levies, to search the village of Tappeh Shuankarah for arms. He returned with eight prisoners and nine rifles. September 16th.

The columns returned to Sulaimani, and Chemchemal respectively, next day. September 18th.

Orders were received from Air Headquarters for a column to march from Sulaimani to the area known as Shar Bazher, the object being to cause Sheikh Mahmud to evacuate the passes on the Kani Manga and so let the Jaf tribes through, without paying taxes to him. The column consisting of:— Column to Shar Bazher.

Two Companies, 2nd Battalion, Iraq Levies,
Three Companies, Iraq Army, and
W/T.

September 18th.

left Sulaimani with ten days' supplies on September 20th, and marched through Chaortah and Dehgalah to Gola, on the northern end of the Penjvin Plain, arriving September 23rd. Sheikh Mahmud evacuated the Kani Manga and retired into Persia, the Jaf marched south and the column came back to Sulaimani on September 30th. Very little opposi- September 20th. September 30th.

October 5th.

tion was encountered. On October 5th one company, 2nd Battalion, Levies, marched to Halebja with a company of the Iraq Army and relieved the two Levy Cavalry Regiments which went to Chemchemal.

In October another column left Sulaimani, by orders received from Air Headquarters, to go through the area east of the main road to Sulaimani, and south of the Qaradagh, with the object of dealing with certain villages in the area which had been in revolt against the Government. This area was inhabited by supporters of Kerim Futteh Beg, and had given a good deal of trouble.

The column consisted of the 2nd Battalion, Iraq Levies (less 2 Companies) under Captain R. Merry, and the 2nd Battalion, Iraq Army, under Colonel Hassan Hilmi.

October 15th. October 17th. October 18th. Column in Chemchemal area.

This column left Sulaimani on October 15th, arrived Chemchemal on October 17th, where Captain Growdon with some Police joined the column. Left Chemchemal on October 18th and marched through Shirdarreh, Hafta Chasmeh, Qadr Karm, Ustakhadr, and back to Chemchemal, destroying villages of those implicated in the rebellion against the Government, and rounding up stock.

October 18th. October 28th.

It then went through the Dar-i-Qeli area, for the same purpose, and arrived back at Sulaimani on October 28th. No opposition was encountered.

While these operations were going on the 2nd Cavalry Regiment took part in a small operation against Qara Anjir, on the road between Kirkuk and Chemchemal. This village had been involved in a number of road robberies and attacks on convoys. An attempt had also been made to wreck the armoured cars. Qara Anjir was surrounded and burnt.

November 27th/28th. Raid on the Ismail Uziri.

On the night of November 27th/28th, one company, 2nd Battalion, Levies, under Captain Foweraker, took part in a raid on the Gomaeh section of the Ismail Uziri. These people had attacked the post on the Azmir Pass in June, when Captain Coffey, Liaison Officer with the Iraq Army, had been killed. Therefore the operation was chiefly carried out by Iraq Army troops, the company of Levies being held in reserve. It was a pitch dark night and raining hard, but the objective, a large camp of Ismail Uziri, was surrounded at dawn. The men of the tribe fled up the slopes of Pir Magrun; some sheep and other animals were captured.

December 10th/11th. Raids of night December 10th/11th.

On the night of December 10th/11th, four columns, one of Iraq Levies, two of Iraq Army, and one of Police, went out of Sulaimani simultaneously to round up adherents of Sheikh Mahmud and local people who were wanted in villages within a radius of about five miles. Each column had an objective, that of the Levies being Werawar. It was a bad night, raining hard, and blowing a gale. Fifty-three prisoners were brought in, who were combed out by the Police, and only those known or suspected were detained. Twenty-one rifles were also brought in. Only one column, two squadrons of the 2nd Cavalry Regiment, Iraq Army, met with opposition, and the enemy fled up the Kaiwan Pass pursued by two troops, and got away.

This was the last operation of the year. The winter rains and snow came on in earnest, and the country became impassable.

Two British Officers in the Levies lost their lives this year in aeroplane accidents. One, Lieut. J. E. Griffith, The Dorsetshire Regiment, 4th Battalion, Iraq Levies, was killed in a crash on landing at Kani Utman on May 4th; the other was Lieut. G. D. E. Heather, The Loyal Regiment, 2nd Levy Cavalry, who was killed with Flight Officer M. G. Penny, No. 30 (B) Squadron, when the aeroplane crashed on the Qaradagh about July 8th; they were not found for some days.

May 4th.

July 8th.

Organization of Levy Medical Services.

It was during this year that the Medical Branch of the Levies was completely organized. Colonel Sanderson, P.M.O. of the Levies, had already done a good deal of work during the period of change and unsettled situations of 1922–23–24, especially during the Rowanduz operations of 1923, when the force was operating in the worst possible weather conditions and had practically no sickness. He then accepted an appointment at home and Lieut.-Colonel D. S. Skelton, D.S.O., succeeded him on 18th April. He brought the whole medical side up to a state of great efficiency which was never lost; the Pack Ambulance was trained and equipped, and the arrangements for looking after the families and dependants of serving men were put on a regular footing. This was in accordance with the terms made with the Assyrian Chiefs in 1922. In addition he started the first of the First Aid Classes held under the auspices of the St. John's Ambulance Association, in which over sixty men were awarded certificates up to date and a certain number progressed as far as getting the medallions and labels. The efficiency of the practical and clerical work of the Medical Branch was very greatly due to the assistant and sub-assistant surgeons of the Indian Medical Service, and to the clerical side of the unit. They gained several honours and rewards during their service with the Levies, as is related in this History.

First Aid Classes started.

CHAPTER IX

1926

1926.

IN January Lieut.-Colonel Cameron returned from leave. Lieut.-Colonel Browne handed over command of Sulaimani to him and, on April 29th, took over command of the Levies from Colonel Dobbin, who returned to England to take command of his Regiment.

This year saw the Levies on continuous operations in the Sulaimani area, and also saw the general cutting down started, and brought into effect early the next year.

Changes of Station in the year.

Arrangements had been made for various changes of station to take place about May. The two Cavalry Regiments changed stations, the 1st now being at Arbil with a squadron at Koi Sanjak, and the 2nd at Kirkuk with a detachment at Chemchemal. The 3rd Battalion, who were at Aqra with a detachment of two companies at Amadiya, marched to Sulaimani, leaving Aqra on May 19th to relieve the 2nd Battalion, which had been on operations almost continuously since the middle of 1924. They were now joined on the march by one company from Amadiya, the other followed the Battalion. The 2nd Battalion, after relief, went to Diana Camp complete, the detachments at Koi Sanjak and Chemchemal being taken over by the Cavalry Regiments. On relief of Koi Sanjak on May 31st, Captain Fry, with one company, 2nd Battalion, went to Diana and all the Assyrian families at Koi Sanjak to Mosul. The 4th Battalion left Kani Utman, which from now ceased to be a station, and went with their families to Mosul, sending a detachment of one company to Sheraman, twelve miles west of Aqra, chosen as a camp site, on account of the amount of malaria at Aqra. The buildings at Kani Utman remained for the present, but were later used to build the new barracks at Diana. It arrived there on 20th May. The Machine-Gun Company had Headquarters and one section at Rowanduz and sections at Kani Utman, Sulaimani and Chemchemal. The exchange of the 3rd and 2nd Battalion was made a little complicated, because Colonel Cameron was with his Battalion on operations and was in command directly under Air Headquarters. It was therefore decided that Colonel Cameron should continue in command of

May 19th.

May 20th.

the present operation as long as it lasted; but that the relief of one Battalion by the other in the column should take place at once. Colonel Barke had pushed on to Sulaimani with three companies of the 3rd Battalion, and on arriving at Sulaimani on June 6th he marched with two companies to Ab-i-Tanjero Bridge on June 9th and relieved the two companies of the 2nd Battalion, who returned to Sulaimani on 12th June. The operations of the year will now be described. They began early in spite of the weather.

June 6th.
June 9th.
June 12th.

On March 3rd Colonel Cameron left Sulaimani with one company, 2nd Battalion, Levies, and two squadrons of Cavalry, and one company of Infantry of the Iraq Army, to deal with Kurds reported to be in the Qaradagh area. The Tanjero River was full, and had to be crossed, and the country water-logged. Attempts were made by rapid moves of the Cavalry, supported by aeroplanes, to catch the enemy who were reported to be in Jaafaran and Zerguez, but none was found; the column returned on March 5th.

March 3rd.
Column to the Qaradagh.
March 5th.

On March 9th, owing to the report that Kerim Futteh Beg and Saber were in the Qaradagh collecting taxes, Cameron left with the intention of trying to catch him on the Tanjero River line. He went himself to Ab-i-Tanjero Bridge with one company, 2nd Battalion, Levies, the 2nd Iraq Cavalry Regiment (less one squadron) and one company, 4th Battalion, Iraq Army, and sent Captain F. R. Grimwood, D.S.O., with one company, 2nd Battalion, Levies, and one company, 4th Battalion, Iraq Army, to Qaradagh. The line of the Tanjero River was piquetted by Cameron's Column, the Police, and the Halebja Garrison, which consisted of two companies, Iraq Army. The Cavalry were used for patrolling the villages in the Tanjero Valley, and to pursue if required. Grimwood was to try and drive Kerim Futteh Beg and his men up against the line of the Tanjero where it formed a loop where it turns from south-east to south.

March 9th.
Column to the Qaradagh against Kerim Futteh Beg.
March 9th.

On March 10th was very heavy rain and the river rose three feet. On March 12th Grimwood marched from Qaradagh, passed Nauti and up the Faqra Pass. Here the enemy were met, and a short fight ensued, in which R.K. Zia Nannoo with some fifteen men held off a Kurdish attack, while R.K. Gewergis Shabo counter-attacked under cover of his two Lewis guns. The Kurds scattered and fled in all directions. Two dead were found and twelve rifles picked up. No further enemy were seen, and during the night of March 13th, the river having dropped, Kerim Futteh Beg and his men crossed it and escaped.

March 10th.
March 12th.
Action at the Faqra Pass.
March 13th.

On March 14th Grimwood joined Cameron at Seara, and on 15th the column returned to Sulaimani.

March 14th.
March 15th.

In June Lieut.-Colonel Cameron left Sulaimani in command of a column with the object of helping the migrating Jaf tribes to move from their winter pastures in Iraq into Persia, without paying the tax levied on them by Sheikh Mahmud.

The Column consisted of :—

2nd Battalion, Iraq Levies (less 2 Companies).
Two Squadrons, 2nd Cavalry Regiment, Iraq Army.
One Company, 4th Battalion, Iraq Army.

Column to the Penjvin area.

A total of 679 personnel, and including a large convoy of 672 animals.

Two Flights No. 1 (Fighter) Squadron, and one Flight No. 30 (Bomber) Squadron were detailed for duty with the column, the general direction of whose operations was from Baghdad.

June 8th. June 10th. June 11th.

The column left Sulaimani on June 8th and went to Ab-i-Tanjero Bridge, where the two companies of the 3rd Battalion arrived and relieved the companies, 2nd Battalion, who went back to Sulaimani. The column marched to a camp at Sarao and then to Kaolas, arriving on June 11th. The whole plain was covered with Jaf, and it was reported that Sheikh Mahmud was raising a lashkar. The first intimation of hostilities was, that a Kurd approached a cavalry standing patrol, and fired on and killed the horse of the sergeant.

June 12th.

On the 12th June the column advanced into the hills and, from 08.00, the right flank guard was engaged, having one man of the Levies wounded. The column crossed the Qarachulan Chai (also called the Ab-i-Douleh), and camped on a small flat piece of ground two miles north of Nalparaz. Reconnaissances found nothing; but information gave Sheikh Mahmud with eight hundred men on the Kani Manga, between the column and Penjvin.

June 13th. June 14th.

On June 13th an Iraq Army reconnaissance under Captain Rich, Q.V.O.C. of Guides, Infantry, met Kurds two miles from camp, and attacked and drove them off, losing one wounded. Enemy here seen in groups on the Kani Manga. On the 14th June, Sheikh Mahmud was reported to have been reinforced, and at 10.00 a.m. he began sniping, followed by an attack on the camp which lasted for four hours, down the spurs of Kani Manga and from the east side. At about 14.00 hours they succeeded in occupying a point on the left flank which enfiladed several defended localities. Lieut. H. M. Curteis, H.L.I., with one platoon of the 3rd Battalion Levies, counter-attacked and drove back the Kurds, killing the leader and two others. On the east (right) flank, the Kurds penetrated the line but were held up. The thickness of the woods prevented support by fire and air action. Three men in the column were wounded. Lieut. Curteis and R.K. Barkhu Hormis were mentioned for

Action of the Kani Manga.

good work and Pte. Khaninia Yakub for gallantry in action. During this day an aeroplane crashed behind the enemy's lines, and the pilot and mechanic were captured and held prisoners for some time, but well-treated.

June 15th and 16th, the column remained in position, carrying out reconnaissance, while the aircraft bombed and machine-gunned the crest of Kani Manga. Sheikh Mahmud was reported to have been further reinforced partly by the Jaf and partly by Persian Kurds, bringing the strength of his force to about two thousand. June 15th. June 16th.

On June 17th a combined patrol of Levies and Iraq Army initiated by the Native Officer in command suddenly attacked enemy snipers on the south side of the Qarachulan River and drove them off. One Iraq Army soldier was wounded. On June 18th, the enemy sniped the camp and were bombed and machine-gunned by No. 30 Squadron. June 17th. June 18th.

They made a final effort next morning. Fire was opened all round at 02.00 hours and the right flank was attacked at 4.30 a.m., but this attack was beaten off, and after being machine-gunned by both the R.A.F. Squadrons the enemy's fire slackened. Sheikh Mahmud's force retreated that morning. At 14.00 Cameron sent Barke forward with one company Levies and one company Iraq Army to reconnoitre Kani Manga; they found it clear. Cameron at once moved on with the column and occupied the top of the hills. Penjvin was occupied on June 21st after slight opposition. June 19th. June 20th. June 21st. Penjvin occupied.

The column left Penjvin on June 25th and arrived at Kurmal on the 26th, where it was intended to meet Captain Brodie of the Iraq Army with one company. He had pushed on ahead, and met Kurds in the place. A fight ensued in which two of Captain Brodie's party were killed and himself and three others wounded. The arrival of the column saved the situation and the enemy retreated, leaving three dead. June 25th.

From June 27th to July 9th the column remained in Kurmal, taking action against rebels and burning crops. They were back in Sulaimani on July 31st. A company of the Iraq Army was left to garrison Kurmal. June 27th. July 9th. July 31st.

While these operations were taking place the 1st Levy Cavalry Regiment had done some most useful work round the area between Kirkuk and the Basian Pass. Up to May the 1st Levy Cavalry Regiment had been at Arbil, and the 2nd at Kirkuk, with a detachment at Koi Sanjak.

In May the two Regiments changed stations; the 1st Regiment arrived at Kirkuk from Arbil on 18th April; the 2nd Regiment left Kirkuk and arrived at Arbil on 23rd April. The detachment at Koi Sanjak was now found by the 2nd Cavalry Regiment at Arbil, and the 1st Levy Cavalry Regiment at Kirkuk put a squadron at Chemchemal and relieved a com- April 18th. April 23rd.

April 18th.
April 23rd.

pany of the 1st (Marsh Arab) Battalion there which joined the Levies at Sulaimani.

June 9th.
June 10th.
June 11th.
June 12th.
June 14th.
June 17th.
Levy Cavalry operations, June 18th, against Kerim Futteh Beg.
June 19th.
June 22nd.
June 23rd.
June 24th.
June 24th.
Death of Kerim Fetteh Beg.

During the night of June 9th, information was received that Kerim Futteh Beg was at Muzzafar. " C " Squadron, 1st Levy Cavalry, left early next morning, made a twenty-eight mile march and surrounded Muzzafar, and found him gone. They visited villages around on 11th and 12th and did the sixty miles back to Kirkuk in two marches, getting in on June 14th. On June 17th he was reported at Dar-i-Qeli, fifty miles from Kirkuk. Lieut.-Colonel L. Alexander left Kirkuk that night with " A " Squadron, joining " B " Squadron under Captain Kinnaird at Chemchemal at 04.00 on June 18th. He intended a continued operation, but at 8.30 a.m. a message was dropped to say that an aeroplane of No. 30 Squadron had come down at Memlaha, thirty miles away, in enemy country. Captain Kinnaird left at once with " B " Squadron, and brought back the pilot and mechanic, covering the sixty miles in twelve hours. He got in at 9 p.m. on 18th June. Kerim Futteh Beg being reported at Mortaka, Colonel Alexander left next day, June 19th, and tried to catch him by a rapid march there. This he evaded, and the squadron went back to Kirkuk. " B " Squadron, from Chemchemal, reconnoitred the area on June 23rd and 24th, and he was located near Qaratumar, and at 01.00 on the 29th June, Colonel Alexander marched to surround the place. The result of seeing him chased about encouraged the people of the area to resist this man, who was simply living by robbery and terrorism. Colonel Alexander not finding him at Qaratumar, withdrew his squadron and hid in the wadys near by, and then left secretly and made for Mortaka, thinking Kerim Futteh Beg would be there. However, he left Mortaka before the squadron arrived, and got back to Qaratumar, and here the villagers, not knowing the squadron had gone, and relying on it to back them up, shot at Kerim Futteh Beg's party as he came up, and gave him a wound from which he died. This had a very great effect on the area.

June 14th.
Column into Qaradagh.
June 16th.

On the return of the column in the Sulaimani area to Sulaimani in July, Colonel Cameron left and took command of his own Battalion at Diana. Colonel Barke, Officer Commanding 3rd Battalion, took over command of the Sulaimani area. Headquarters and the rest of the Battalion arrived on July 1st. In the absence of the column, steps were taken to bring the Qaradagh into order. On June 14th two platoons of the 1st (Marsh Arab) Battalion Levies left Sulaimani and marched through the Gilazarda Pass to Qaradagh, arriving on June 16th. Here they remained holding a position guarding the building of the new Police Post there, until recalled to go back to Baghdad, to rejoin their Battalion which was now to be transferred to the Iraq Army.

The operations in Sulaimani were continuous until the end of the year.

On August 20th Colonel Barke left with Headquarters and one company 3rd Battalion, Iraq Levies, one Sub-Section Machine-Gun Company, and one company 4th Battalion, Iraq Army, on a reconnaissance march through the area on the north side of the Qaradagh from Derbend-i-Bessira to Paikuli Pass. The object was to get a thorough knowledge of the country with a view of being able to control the moves of the Jaf in future. The column went through the area and was back in Sulaimani on August 30th, with no opposition.

August 20th.

Column to Paikuli.

August 30th.

On September 6th Colonel Barke left Sulaimani again with a column composed of Headquarters and two companies 3rd Battalion, Iraq Levies, a section of Machine-Guns, Iraq Levies, and one company, Iraq Army. The object of this column was to be used as necessary in connection with Salah al Dowlah, the Persian pretender, who was known to be in Iraq. However, on arrival at Kurmal on September 8th, Barke was ordered by Air Headquarters to return to Sulaimani, leaving the carrying out of the present objective to an Iraq Army Column. He was back in Sulaimani on September 12th, and was at once ordered off again to the Gola area with a column, with the objective of preventing Sheikh Mahmud taxing the Jaf as they passed south over the Penjvin Plain. In addition he was to co-operate with the Iraq Army Column at Kurmal, which was placed under his orders. He took Headquarters and three companies 3rd Battalion, Iraq Levies, one Machine-Gun Section, Iraq Levies, and two companies 4th Battalion, Iraq Army. A W/T. set went with the column which also had one Flight No. 1 Squadron, and one Flight No. 30 Squadron to co-operate. The column left Sulaimani on September 16th and reached Gola on the 18th. Some slight opposition was encountered near Qizilja, and the Special Service Officer's party was fired on near Penjvin. Negotiations had been going on between Sheikh Mahmud and Air Headquarters on the subject of the two airmen prisoners, Flight Lieutenant Denny and Air Mechanic Hirst, captured in June. They were both ill. By arrangement with Sheikh Mahmud, Captain F. R. Shaw, M.C., Medical Officer, Iraq Levies, went into Sheikh Mahmud's headquarters at Walajia, and saw them, a truce being arranged meantime. Sheikh Mahmud agreed to hand them both over, provided it did not look as if he was being forced to do so. The truce was prolonged, the column returned to Sulaimani on October 6th, and Sheikh Mahmud returned the airmen on October 9th.

September 6th.

Column to Kurmal.

September 12th.

September 16th.

September 18th.

September 21st.

September 23rd.

Column to Gola.

Release of Captured Airmen.

October 6th.

October 9th.

On October 16th, Barke left again to assist the Civil Authorities in collecting fines and taxes from the Jaf and to arrest certain people. He took out two companies, 3rd Battalion, and was joined by one Machine-Gun Section, Iraq Levies, and two companies, 4th Battalion, Iraq Army,

October 16th.

Column to Ab-i-Tanjero.

on the Ab-i-Tanjero. They left Sulaimani on October 16th. One company Iraq Levies joined an Iraq Army Column now under Captain Grant at Paikuli. The column moved through the area, keeping in touch with the Civil Authorities, and after the completion of the work returned on November 10th.

November 10th.

November 28th.

On November 28th, the last column went out before winter closed down. Barke took the same force as usual, two companies, 3rd Battalion, and a Machine-Gun Section of the Levies, and a company of the Iraq Army. This column left on November 28th and went through the Gorakuli and Paikuli Passes to the Diala about Paibaz. Except by the Police, no opposition was met with. The Police Post in Paikuli was sited, and the column was back again in Sulaimani on December 3rd. Winter came on thoroughly two days later and all operations ceased.

Column to the Diala.

One other small affair in which the Levies were involved took place near Rowanduz at the end of the year. Colonel Cameron was in Diana with the 2nd Battalion, where it had been since its extremely active time in Sulaimani.

The Qaimakhan of Rowanduz was at this time Sayid Taha, a leading Kurd put in by the Government. He wished to arrest a Kurdish chief near by called Ahmed Begok, and, by orders from Air Headquarters, Lieut.-Colonel Cameron was ordered to co-operate. The Qaimakhan proposed to Colonel Cameron, that he with his Police should surround Ahmed Begok's village, Musa Khowa, before dawn of December 8th, while Cameron, having marched out the day before, should block all means of retreat north and west. It was a very cold night and much snow. Cameron carried out his part of the plan, but Sayid Taha attacked two hours before he should have, and Ahmed Begok escaped. Cameron had a little fighting with some of the aroused supporters of Ahmed Begok and had one man wounded.

Attempt to arrest Ahmed Begok.

December 8th.

It had been arranged to evacuate Sheraman, and for the company there to come to Mosul. Various reasons were put forward by the Civil Authorities, and finally the High Commissioner decided that there must be a Levy garrison in Aqra. Sheraman was therefore evacuated and the old lines at Aqra repaired, and occupied by a company of the 4th Battalion on December 7th.

December 7th.

1ST Marsh Arab Battalion at BAGHDAD with detachments (temporarily) at SHERGAT and SULAIMANI up to their change over to the IRAQ Army. Remainder of Levies in KURDISTAN areas until they took over defence of HINAIDI, and BAGHDAD guards in 1928.

BOUNDARIES.

TURKEY
PERSIA
SYRIA

MILES
0 25 50 100 150 200

Areas of Levy Operations 1921-1930.

CHAPTER X

1926–1927

THIS year the cutting down of the strength of Levies by amalgamations, or by disbandments, really began. This was in accordance with the policy, that as the Iraq Army units became trained the Levies should disappear. 1926.

In June came orders from Air Headquarters for the amalgamation of the 1st and 2nd Cavalry Regiments. They were located as follows:— Amalgamation of Two Levy Cavalry Regiments.

1st Cavalry Regiment at Arbil with a Squadron at Koi Sanjak. Commanded by Lieut.-Colonel C. R. Terrot, D.S.O.

2nd Cavalry Regiment at Kirkuk with a Squadron at Chemchemal. Commanded by Lieut.-Colonel L. Alexander.

This amalgamation made a reduction of 7 British Officers, 2 British N.C.O.'s, and 350 Native Ranks.

The amalgamation took place as follows:—On June 19th the Iraq Army took over Arbil, which freed Colonel Terrot to get on with the arrangements. June 19th.

On August 14th the Squadron 1st Cavalry Regiment left Koi Sanjak, arriving at Arbil on the 16th, which concentrated the 1st Cavalry Regiment. The amalgamation was completed, on paper, by the third week in August, the officers and men of the Cavalry all proceeding to Kirkuk, where the new 1st/2nd Levy Cavalry Regiment was stationed. All work was completed, and the small office kept open there, closed at Arbil on 5th October. Colonel Alexander took command of the new 1st/2nd Cavalry Regiment, Colonel Terrot proceeded to Mosul and became second-in-command of the Levies. The whole amalgamation went smoothly in spite of having to provide cavalry escorts from Kirkuk to Sulaimani during that period, and of the appearance of raiders in the area. August 14th. August 16th. October 5th.

The following notice appeared in Orders on the completion of the amalgamation:—

"I cannot let the 1st and 2nd Cavalry Regiments disappear as separate units from the Iraq Levies, without expressing my thanks to them for their excellent work in the past, and my great regret at their amalgamation.

" Throughout the last two years they have been constantly engaged in operations in Kurdistan, and have engaged with success.

" Their turn-out, discipline, and horses have been a credit to themselves, and to the whole force, and I wish to thank all Officers, Warrant Officers, and N.C.O.'s and Men, whose hard work has made this result possible."

In July came the orders for the transfer of the 1st Marsh Arab Battalion to the Iraq Army.

1927. January 1st. 1st Marsh Arab Battalion transferred to the Iraq Army.

The date on which this was to be effected was on and from the 1st January, 1927. A Cabinet crisis occurred during the period arrangements were being made and delayed this transfer being completed until 2nd February, 1927.

Two companies of this Battalion were doing duty in the Sulaimani area, and the Battalion was carrying on guard duties in Baghdad. The companies were sent back, and guard duties taken over by Indian Army units, and the Battalion was concentrated at Baghdad West Barracks.

The Ministry of Defence agreed to take over four British Officers, and two British N.C.O.'s, the rest going to other Levy units or leaving. Three British Officers took on.

The most difficult point was that of the Native Officers. The crux was education. The Iraq Army Officer is expected to have sufficient education to do the executive work required in a Battalion. The Officers of the 1st Marsh Arab Battalion had never done anything of the sort. The British Officer had been in executive command and had done all the work required. The Native Officer had been picked for ability in practical work, for tribal reasons, i.e. that he was a person of influence, and hence one the men would follow, or for having distinguished himself in the field. Education played no part. Against that was the obvious point, that if the Iraq Army authorities refused to take the present Native Officers, they would run the risk that the majority of the men would refuse to take on under the Native Officers whom they did not know. However, it was decided they must pass an education test and one was arranged, with an officer of the Levy Headquarters on the board. Out of the seventeen Native Officers, thirteen presented themselves for examination. All failed. One only had more than an elementary knowledge of reading and writing Arabic.

Of the men on 1st February, 1927, 111 transferred at once, and 78 more re-enrolled in the Battalion after transfer.

Captains G. S. H. R. V. DeGaury, M.C., and E. V. Packer went to the Iraq Army with the Battalion, and Captain J. N. Donnellan went to the

Iraq Army as Signalling Officer. Lieut.-Colonel F. J. E. Archer, commanding the Battalion, left the country after handing over and later took over the raising and training of the Body Guard of the Maharajah of Cutch.

The 1st Battalion became the 7th Battalion of the Iraq Army.

At the same time the disbandment of the Levy Pack Battery was taking place. Orders for this to take place had come on 14th December, 1926. At the same time a letter was received from the High Commissioner's Office saying :— 1926. December 14th.

> " In connection with the orders which are being issued separately regarding the disbandment of the Levy Pack Battery, His Excellency wishes to record his deep regret at the forthcoming disappearance of this Unit, of whose smartness and efficiency he has frequently had the most favourable reports. Disbandment of the Levy Pack Battery.
>
> (Sd.) F. E. STAFFORD.
> Financial Secretary to His Excellency
> The High Commissioner for Iraq."

The Air Vice-Marshal associated himself with these expressions of regret.

The date for the disbandment to begin was fixed for 24th January, 1927. Major V. R. Guise, M.C., the Commander of the Battery, returned to duty in England. To effect the reduction in strength in the force caused by the disappearance of this unit, the places of one captain, and four subalterns, due to leave on completion of tour of duty, were not filled up. 1927. January 24th.

Of the British N.C.O.'s B.S.-M. W. Clark went back to duty, and the other three were absorbed in other units of the Levies. Of the Native Assyrian Officers two were absorbed in other Battalions, and one discharged. The Assyrian battery sergeant-major, sergeants, corporals, and other ranks were distributed in the Battalions. Forty-six trained pack leaders went to the 4th Battalion and took the place of forty-six Yezidis, who were discharged. The animals went chiefly to Remounts ; guns and equipment to the Iraq Army.

This disbandment was effected by 28th February, 1927. On the disappearance of the 1st Marsh Arab Battalion and the Pack Battery, the following Special Order of the Day was issued from Levy Headquarters. February 28th.

" Mosul.
29th January, 1927.

Special Order of The Day. January 29th.
By
Colonel-Commandant J. G. Browne, C.M.G., C.B.E., D.S.O.,
Commanding, The Iraq Levies.

" On the occasion of the disbandment of the 1st Battalion (Arab) Iraq Levies, and of the Levy Pack Battery, I wish to express my great regret at their disappearance from the Levy Force, and to convey to them my thanks for the excellent work they have done in the past.

"Their turn-out, discipline and keenness have been a credit to themselves, and to the Force as a whole, and I wish to thank all Officers, Warrant Officers, Non-Commissioned Officers, and Men for the work they have so ably carried out, thus enabling this high standard of efficiency to be maintained."

1927.

The cutting down continued during the year. The next unit taken in hand was the Depot, which was cut down into a much smaller unit under a Commanding Officer, with an Officer under him doing the work of Adjutant and Quartermaster combined, and the Depot divided into four groups:—

Reorganization of the Depot.

A Group. Recruits, Collecting, Training and Distributing.
B ,, Employed Men of the Mosul Garrison.
C ,, Transport and Remounts.
D ,, Garrison and Camp Police.

The Officer Commanding Transport and Remounts, Signalling Officer, and Officer Commanding Machine-Gun Company were also attached to the Depot for administration and discipline.

August 1st.

The modification was effected by August 1st.

With the re-organization of the Depot, it was decided that a Veterinary Officer was no longer needed at Headquarters, and Lieut.-Colonel J. A. B. McGowan left on 30th April.

April 30th.

March 3rd.

Disbandment of the 1st/2nd Levy Cavalry Regiment.

Orders had already come to cease recruiting and re-engaging for the 1st/2nd Levy Cavalry Regiment. The date for disbandment of this unit was finally ordered to be effected, after some correspondence, by December 17th, which meant that its use as a fighting unit would cease on October 31st. The animals of the Regiment went to private purchasers, to Levy units, exchanging good cavalry animals, for those requiring replacing, to the Police and Iraq Army. About eight officers and a number of men joined the Iraq Army, and many joined the Police.

Prior to their disappearance the Regiment, in addition to carrying out some very active and useful operations round Kirkuk area against the rebels there, which will be described later, also won a large number of events at the Baghdad Horse Show and were most successful at polo. The Regimental Team won the Lloyd-Sargon Cup at Mosul, and followed that up by winning the Reid Cup at Baghdad in November, just before their disbandment.

On 15th October the Acting High Commissioner issued the following notice:—

"In connection with the disbandment of the 1st/2nd Cavalry Regiment, which is now proceeding, the Acting High Commissioner wishes to record his deep regret at the disappearance of this Unit. He has frequently received the most favourable reports of the work and efficiency of the Regiment, and is sorry that the dictates of policy do not permit of its retention.

(Sd.) F. E. STAFFORD.
Financial Adviser to H.E. the
High Commissioner for Iraq."

December 17th.

The Regiment ceased to exist on December 17th. Owing to the disbandment of units, orders came from the High Commissioner to effect a reduction of Officers amounting to one lieut.-colonel, two captains, four subalterns, and also one company or squadron sergeant-major, and one quartermaster-sergeant. This was effected.

These disbandments brought the Levies down by the end of the year to a Headquarters, three Battalions, Machine-Gun Company, the Depot, and Medical Branch. Further disbandments were to come in the next year.

1927. March.

In March, 1927, it was decided to bring the activities of Sheikh Mahmud to an end. So far he had always had Penjvin as a place at which he could stay, when it was not occupied temporarily by one of our columns. The next operations had therefore the objective of (*a*) Occupying Penjvin and bringing that area under administrative control, (*b*) In carrying this out the infliction of a severe defeat on Sheikh Mahmud and his following.

Operations against Sheikh Mahmud.

It was comparatively easy to carry out the first part of the objective. This had been done before. To remain there was a little more difficult, and meant much convoy work. But to inflict a defeat was another matter. Since he had held the Kani Manga in June, 1926, Sheikh Mahmud had not attempted to concentrate any force together, and since the battle of the Bazian in 1919, he had not attempted to fight a pitched battle. The only chance was to catch him by surprise, and this was difficult. The report, said to come from Sheikh Mahmud himself, that he knew all that went on in Sulaimani within six hours of its occurrence, was fairly correct.

It was on this account that a method of effecting a surprise was suggested by Colonel Barke, commanding at Sulaimani, and was put into effect for these operations.

The chief means by which information leaked out was through local transport. Levies had their own Transport Company. By cutting down the animal ration to about 6 lb., and further by not collecting the animal ration in Sulaimani, it would be possible for the Levy Column to leave Sulaimani on operations and remain out for six days without making any use of local transport or supplies. After that date they must either get back to Sulaimani, or be met by a convoy. The plan was therefore, shortly, for one Column called Defcol to leave on a certain date, drawing supplies and escort in the ordinary way, and march on Penjvin. It was no good trying to disguise its objective ; that always leaked out. After this Column had left, the Levy Column called Levcol to leave with its own transport, and march by a different route. As regards supplies, six days for men and animals was sent up direct from Baghdad to No. 30 Squadron at Sulaimani and held in their lines until drawn by the Levies just before leaving. Being British rations they excited no comment, it being thought they were for the R.A.F. It was hoped that Sheikh Mahmud might oppose the first Column with considerable forces and, while engaged, might be caught by the Levy Column. It was almost certain that if the first Column advanced on Penjvin from the direction of Kaolas, that the force of Sheikh Mahmud would occupy the Kani Manga, and this gave the chance that, if the Levy Column came unexpectedly on his rear through Waliawa and Qizilja, it might cut off a good part of his force from escaping into Persia. This would be on, or about, the third day after the Levy Column had left Sulaimani. A Supply Column called Supcol must therefore leave Sulaimani, so as to be in Penjvin not later than the sixth day. The departure of this Supply Column would be connected, it was hoped, with the first Column which had left.

The Air Vice-Marshal Sir Edward Ellington, K.C.B., C.M.G., C.B.E., approved of the scheme, and ordered Colonel Browne, Commandant of the Levies, to be in command of the whole operation. Headquarters of the operation assembled in Sulaimani on April 12th.

April 12th.

April 13th.

Kurds attack Kani Spika.

On the 13th April, however, Sheikh Mahmud's party became active, attacking Kani Spika, four miles from Sulaimani, at about 6 p.m., and Kosta Chem. The latter defended itself successfully. Kani Spika was captured and looted, and several people killed. The news of the attack only reached Sulaimani at 7.30 p.m., a company of Levies left at once under Captains C. E. P. Hooker and R. Edwards, but the enemy had gone.

April 19th.

On April 19th Defcol left and went through Arbat to Sarao. This was a column entirely composed of Iraq Army troops and Police under the command of Major I. Clayton.

April 20th.

On April 20th Levcol drew their supplies, held in No. 30 Squadron lines, and marched. No one except the Headquarters and Colonel Barke and his staff knew they were to move until they left.

April 21st.

April 22nd.

There was heavy rain on the night of April 21st and all streams and rivers were flooded. It cleared about midday and Defcol marching to Nalparaz were attacked on their right flank, and had about ten casualties. They camped at Nalparaz, the Ab-i-Douleh River was in flood, and Kurds held the Kani Manga beyond it. Next morning Defcol crossed the Ab-i-Douleh, the Kurds on their right flank and rear being held off by rear-guards and by the action of the No. 30 Squadron, who machine-gunned and bombed them, when they tried to come on, and also attacked the Kurds on top of the Kani Manga. April 23rd. Action at Nalparaz.

Barke's Column, Levcol, had marched rapidly and occupied Penjvin on the evening of April 22nd, at 6 p.m. One of his piquets on the south side of the river at Waliawa had been cut off by the rise of the river and had to return to Sulaimani. April 22nd. Penjvin occupied.

At 6.30 a.m. on April 23rd, he left Penjvin, and advanced up the Kani Manga on the rear of the Kurds opposing the march of Defcol. The surprise was effected, but the enemy fled at once from the position, leaving some grazing horses and two dead behind. As they went they were caught under machine-gun fire by Lieut. N. Patterson, Officer Commanding Machine-Gun Company. Defcol got to the top with only a few shots fired at the start of the action, when No. 30 Squadron bombed along the top of the Kani Manga. April 23rd. Action on Kani Manga.

Both Columns reached Penjvin that night. The Supply Column under Captain Hooker arrived next day at Nalparaz. Captain Baddiley with one and a half companies of Levies occupied the top of Kani Manga and there was no opposition. Penjvin defences were sited and work started, and a new Mudir installed. The punishment of the villages, which had taken part in the attack on Defcol, was carried out by No. 30 Squadron by bombing action. April 24th.

On April 27th, Colonel Barke left Penjvin with part of the force, partly Levies and partly Iraq Army, for Sulaimani. He was obliged to make a long and rapid march in heavy rain, so as to get across the river at Waliawa before it became too deep. This he just succeeded in doing. April 27th.

On May 2nd, Force Headquarters with one company 3rd Battalion, and one section of Machine-Guns started back for Sulaimani. The same day a large Supply Column, escorted by one company 4th Battalion, under Captain Edwards, was attacked on its way to Sulaimani at Waliawa. The action went on for some four hours, but the attack was beaten off and the Levies chased the Kurds for some distance, killing six, and having C.Q.M.S. Baitu, 4th Battalion, and one other man wounded, one animal hit. May 2nd. Action at Waliawa.

This Column, called Nedcol on account of its numbers of donkey transport animals, and the Column with Force Headquarters called Hookcol, passed each other on the Tarnier Dagh on May 3rd. May 3rd.

On the morning of May 4th Hookcol, which had camped at Waliawa, was attacked while leaving. The attack was limited to long-range fire and ended in the Kurds retreating, leaving two dead behind. One incident was, that the wounded C.Q.M.S. Baitu escaped from the ambulance and joined in the fight.

This ended the operations.

The Levy Officers and men mentioned for good work on these operations were Lieut.-Colonel C. R. Barke, C.B.E., Major E. T. Horner, M.C., Major R. H. L. Fink, M.C., Lieut. E. G. Buckley, Rab Khamshi Eshu Saper, C.Q.M.S. Baitu Mako, and Corporal Barkhu Bobo.

May 11th. 4th Battalion at Sulaimani.

Lieut.-Colonel J. M. Gillatt, D.S.O., Commanding the 4th Battalion, Iraq Levies, left Mosul for Sulaimani with his Battalion, and on May 11th took over command of Sulaimani from Lieut.-Colonel Barke who marched his Battalion back to Mosul. The operations against Sheikh Mahmud were gradually dying down; but each convoy to feed the Iraq Army garrison at Penjvin required a strong escort, which was provided by the 4th Battalion Levies. (May 20th.) On May 20th, Lieut. Hart, in command of one such escort, met Kurds two miles from Chingina, and in a small action inflicted upon them a loss of two killed, and one wounded. Shortly after this, Sheikh Mahmud made terms with the Government.

Some of his supporters however attempted to continue the rebellion. Saber and Abdullah, the sons of Kerim Futteh Beg, went to their old area near the Bazian Pass, and again began robberies and murders.

At the request of the Administrative Inspector, Lieut.-Colonel Alexander took action at once with the 1st/2nd Cavalry Regiment.

Levy Cavalry operations against Kurds in the Chemchemal district.

On May 25th he left Kirkuk and on May 29th and 30th, assisted by two squadrons Iraq Army Cavalry, placed under his command, he drove northwards with five squadrons in line up the Qaradagh and Baranand Dagh to a line north of the Bazian Pass. At Takiyah Kullah, a party of rebels attempting to escape from Major S. Fosdick's Squadron ran into that commanded by Lieut. T. C. Hobbs, and left one wounded man behind.

On the way back to Kirkuk, Lieut. Hobbs' Squadron was fired on at Bashbulaq, and rebels were seen lining a ridge in front of that place. He drew swords, and charged in extended order, and the rebels after firing (May 30th.) hard at the horsemen for a few seconds, fled in all directions, some going into a garden in Bashbulaq. Hobbs galloped on into the garden, and here heavy firing ensued until well after dark. Three dead rebels were picked up and three more captured in the garden. It was too dark to see where the rest had gone, for pursuit. The village was burnt.

For this action and his other work on these operations Lieut. Hobbs received the Military Cross.

Saber still continued in the area with a band of about seventy men, (June 14th.) and on 14th June, Colonel Alexander went after him again with his regiment and drove the area of Derbend-i-Bassirah. He met no opposition, but Saber and his band left the area and went back to Persia. The 1st/2nd Regiment returned to Kirkuk on June 24th. June 24th.

One other operation was carried out this year. This was against Sheikh Ahmed of Barzan. The object was the occupation of Barzan, and to bring that area under administrative control. Two Levy Columns carried out this operation, one called Dicol under Lieut.-Colonel Cameron consisting of one company, 2nd Battalion Levies, and one Machine-Gun Section marched from Diana; the other consisting of one company, 2nd Battalion Levies, and one Section Machine-Guns, called Aqcol, assembled at Aqra. No. 6 Squadron, Royal Air Force, co-operated. Captain Littledale, Inspecting Officer of the Police, Mosul, accompanied Aqcol. Colonel Commandant J. G. Browne commanded the operation. Operation against Barzan.

The operation began on June 12th. On the 13th it was known the Sheikh would offer no resistance. However the columns marched to Barzan, a Police Post was sited there. A large rowing-boat was built and placed at the Civil Ferry at Barzan. A bridge-head was sited at the Civil Ferry, to be held by one company of Levies under Captain Fry, M.C.; but as there was no possible landing ground there, it was in August moved to Billeh Camp, where a landing ground is now established. The Levy Columns returned on June 19th. Major Horner's idea of having the large boat made on the spot for transporting troops across the river, greatly assisted the crossing. This boat, named after him "The Johnny Horner," remained a feature of that part of the river until swept away in a flood in 1931. June 12th. June 19th.

This operation ended the active service of the Levies in Iraq, except as garrison troops. With the Southern Desert fighting in 1928 and 1929 they had no concern. With the disappearance of Sheikh Mahmud in 1927, as one member of the Force put it, "the Levies lost their electric hare." On his short appearance again in 1930 and 1931, they were only used for garrison and escort duties. The last years of their service was as a reliable reserve, to be used only in emergency, and, as such, they were useful. A further stage in the disappearance of the Levies now took place. Orders came for the disbandment of the 4th Battalion on November 19th, to be completed by April 30th, 1928. November 19th.

CHAPTER XI

1928

1928.

THE year 1928 opened with a very sad occurrence when Lieut. S. G. Haserick, K.O.Y.L.I., serving in the 2nd Battalion, was killed in an aeroplane crash at Baghdad on January 9th. The pilot of the machine, P/O L. E. R. Fisher, No. 6 (A.C.) Squadron, was killed at the same time.

At the beginning of 1928, the Levies were situated as follows :—

Location of Levies in 1928.

Headquarters, the reconstructed Depot, Transport, Remounts and Levy Pay Office were all at Mosul.

The 2nd Battalion at Diana with a detachment of one Company at Billeh.

The 3rd Battalion at Mosul with one Company at Bebadi.

The 4th Battalion at Sulaimani with a detachment holding Tasluja.

On January 1st the detachment 3rd Battalion at Bebadi handed over to a detachment of the Iraq Army and reached Mosul on February 5th.

February 3rd.

Disbandment of the 4th Battalion.

The orders for the disbandment of the 4th Battalion having gone out, they left Sulaimani on February 3rd, handing over as much as possible in stores to the Iraq Army. This left no Levy troops in Sulaimani. They had a very bad wet march to Mosul, arriving February 22nd. Arrangements were made to discharge certain officers and men from both the other Battalions and fill up their places with the 4th Battalion personnel. All the remaining Yezidis, who were in the 4th Battalion Transport, were discharged. The greatest difficulty to adjust was the case of the Rab Tremma, or senior officer, in the 4th Battalion, Zia Shemsdin. Being son of Malik Shemsdin of the Lower Tiari, he was a man of some standing, and a great power among his own people. But the two Rab Tremmas of the other Battalions were both men of influence, Rab Tremma Daniel Ismail, and Rab Tremma Yakub Ismail, sons of Malik Ismail of the Upper Tiari. The contention of the supporters of Zia Shemsdin was that it was not fair to have two Upper Tiari Rab Tremmas, and no Lower Tiari one, when the Lower Tiari men exceeded the Upper Tiari in providing numbers in Levies. However, the Colonel Commandant decided, on grounds of efficiency, and both the sons of Malik Ismail remained. Zia Shemsdin was given the offer to come in at a lower rank in senior Rab Emma, but declined. In February the disbandment which was going on suddenly had to be checked, as the Barzanis under Sheikh Ahmed seemed likely to make trouble. As soon as this had quietened down and the disbandment was again continued, the Akhwan trouble in the Southern Desert broke out. On account of this the Air Vice-Marshal gave orders for the two remaining Battalions each to retain one hundred other ranks over establishment for the present. February 25th.

The Battalion ceased to exist on March 1st. March 1st.

The following Special Order of the Day was issued:—

> "Owing to the financial stringency at present existing, and to the urgent need for economy in the expenditure of British Imperial Funds in Iraq, it has been decided to reduce the Assyrian Battalions of the Levy Force from three to two.
>
> "The disappearance of the Battalion has caused the deepest regret, not only to myself, but also to His Excellency the High Commissioner for Iraq, and to Air Vice-Marshal Sir E. L. Ellington, K.C.B., C.M.G., C.B.E.
>
> "I take this opportunity of thanking All Ranks of the 2nd and 3rd Battalions, on the change of name of 1st and 2nd Assyrian Battalions, and to the 4th Battalion on their disbandment, for the magnificent work they have done with never failing readiness and willingness, for the past six years.
>
> "I now look to the 1st and 2nd Assyrian Battalions to shoulder the increased duties and difficulties consequent on the reduction of the Force, and to uphold the traditions of smartness, gallantry, and efficiency, so well established by the Iraq Levies in the past.
>
> J. G. Browne.
> Colonel Commandant,
> Commanding Iraq Levies."

As a memorial of this Battalion, whose colour was black, one of the colours of the flag of Iraq, the other two Battalions wore black hose tops to their stockings, shown over the top of the puttees.

Many of the discharged men and some of the Officers went to the Police and some of the men to the Iraq Army. Lieut.-Colonel Gillatt returned to England and other British Officers also did so, or were absorbed.

Authority was received in February that the strength of Levies was to be:—

Headquarters.
Two Assyrian Battalions.
Two Machine-Gun Sections.
Transport and Remounts.

The two Battalions were re-named, the 2nd Battalion becoming the 1st Assyrian Battalion and the 3rd Battalion becoming the 2nd Assyrian Battalion. The Machine-Gun Company was reduced to two Sections by February 29th.

February 29th.

Reorganization of the Staff.

Some reorganization of the Staff due to the decrease in the Force had already taken place, G.S.O. II had gone, and the medical staff cut down from three British Officers to one. There still remained a very efficient body of Sub-Assistant Surgeons of the Indian Medical Service who carried on the medical work in the out-stations. Levy Headquarters offices had to be cut down. For eight years it had been carrying out its duties under the able hands of R.S.M. Walker and his assistants. Some of these had now to go, as the R.A.F. took over certain of the administrative work. Sergeant-Major Walker continued to carry on, with a reduced staff of Indian and Native clerks.

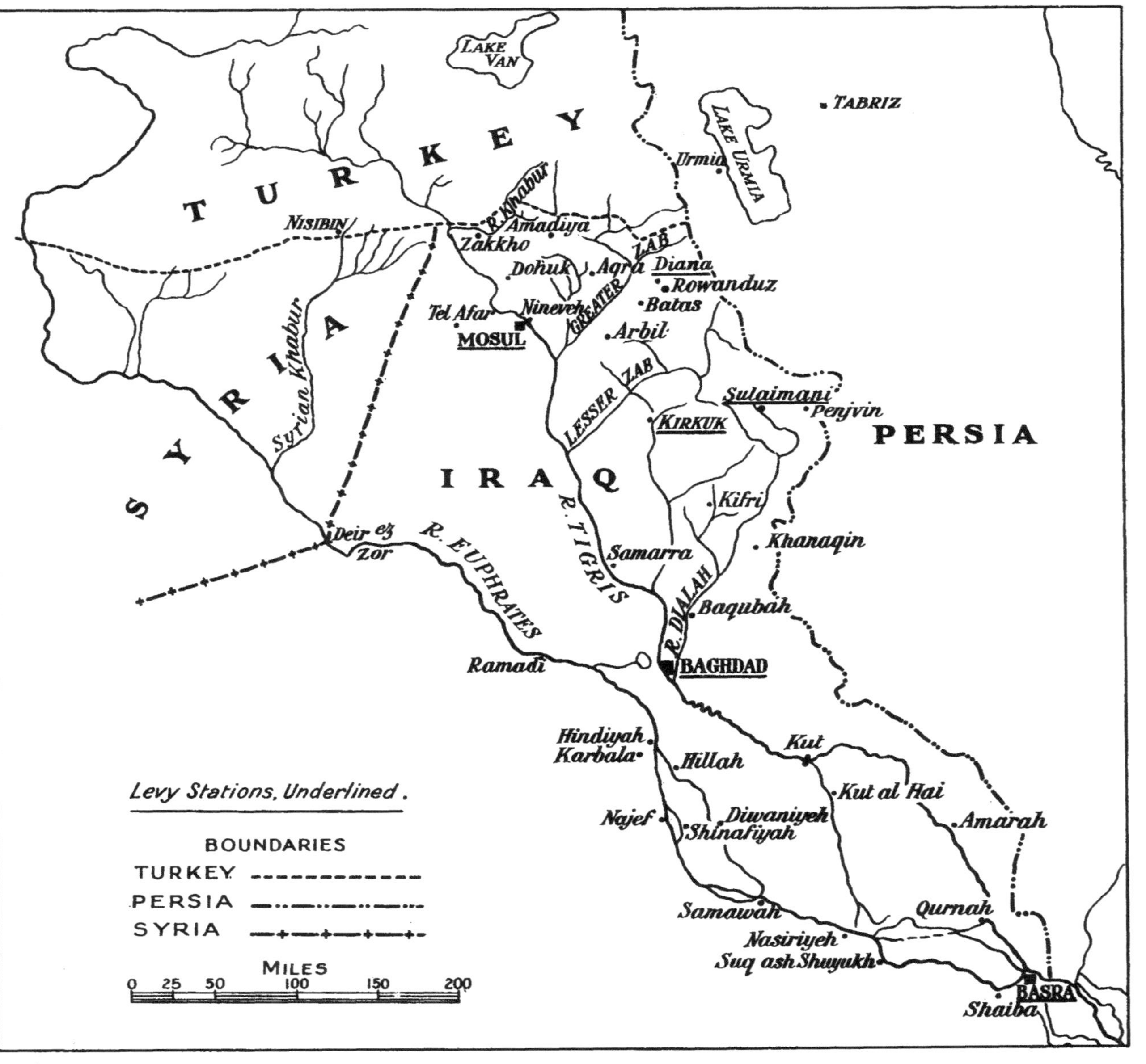

Position of Levy Troops just before the Force ended.

CHAPTER XII

1928–1932

ON July 1st the Iraq Levies passed from the control of the Colonial Office to that of the Royal Air Force, and a new distribution of troops and a reorganization took place, as the Royal Air Force took over Pay, Ordnance, and all administrative work of the Levies.

1928. July 1st. Levies transferred from Colonial Office to the Royal Air Force.

Prior to this taking place, it had been decided that there must still be Levy troops in Sulaimani, so that in May three companies left Mosul, arriving at Sulaimani the same month. They were still billeted in the town, but a site was chosen for a camp near Sheikh Mahmud's pool, and later complete barracks were built with Levy labour. Battalion Headquarters remained for the present in Mosul and finally arrived in Sulaimani on October 1st.

Levy Headquarters move to Baghdad.

The Air Vice-Marshal then ordered Levy Headquarters to leave Mosul for Baghdad. This was partly for economy, as they were to be in the Hinaidi Cantonment, partly for convenience of administration by Air Headquarters, and partly because now the last Battalion of the Indian Army was to go, the Levies were to take over the guards on the High Commissioner's House, the Air Vice-Marshal's House, and certain guards on the aerodrome, as well as to provide numerous escorts.

The following arrangements were therefore made. Levy Pay Office closed down at Mosul, and went to Baghdad on 29th June. The Levy Guard at Mosul aerodrome handed over to the Iraq Army on September 27th who took on this duty.

October 13th.

The Ordnance was closed down in Mosul, and all equipment, not taken down, was sold locally. On October 13th part of Levy Headquarters, Machine-Gun Company, The Depot, Pack Ambulance, Transport, Remounts, and all officers' horses and grooms left by march route for Kirkuk. The Transport and Remounts remained at Kirkuk in the old No. 30 Squadron lines, the rest entrained for Baghdad, arriving on October 25th. Two companies of the 2nd Assyrian Battalion under Captain Aldworth went to Baghdad for guard and escort duties and took over from the 3rd/5th Mahrattas on October 18th. On October 29th, Levy Headquarters finally

October 25th. October 29th.

left their old Headquarters in Mosul, in the street known as "Stink Alley," and opened up at Hinaidi on October 30th, after almost exactly six years' stay in Mosul.

1929. December 1st. On the 1st December, 1929, the Depot finally closed down at Hinaidi, and Lieut.-Colonel Alexander went to Kirkuk and took over command of the Transport and Remounts.

The D.A.Q.M.G. disappeared from the Staff, now that Air Headquarters did all work hitherto done by him, and the Staff consisted of a Brigade-Major and a Staff Captain, later reduced to Staff Officer I and Staff Officer II.

August 2nd. On 2nd August the Levies sustained a great loss in the death of Lieut. M. H. Wallace, H.L.I., who had been Staff Captain for two years. He had just finished his tour of five years and was leaving for England, when he was seized with a severe attack of cerebral malaria and died.

1930. Sulaimani Election riots. On September the 6th, 1930, occurred the election riots in Sulaimani, where the crowd got out of hand and the Iraq Army were called in to restore order. Following this, Sheikh Mahmud suddenly reappeared from Persia, for his third, and probably last appearance, as head of a rebellion against the Government, and attacked Penjvin.

November 3rd. Sheikh Mahmud's Third Rebellion. On November 3rd, by order of the Air Vice-Marshal, Brigadier Browne, Commanding the Levies, proceeded at once to Sulaimani and took over command there, put the defences in order, and issued a temporary defence scheme. The Transport Company under Lieut.-Colonel Alexander marched from Kirkuk to Sulaimani, arriving November 6th. This made the Levies mobile, if required. By the end of November the attack on Penjvin had failed and matters quietened down. The Iraq Army took over the defences of Sulaimani, and advanced Levy Headquarters, which had been established there, returned to Baghdad. For the rest of Sheikh Mahmud's rebellion the Levies were held in reserve. They occupied part of the defences of Sulaimani, provided parties to load up aeroplanes during operations, and so provided a small, if unnoticed quota, to the success of the operations.

1931. In February, 1931, the Transport and Machine-Gun Companies ceased to exist as separate units, both being divided between the two Assyrian Battalions. Kirkuk was occupied by a platoon from the 2nd Battalion to guard the Wireless and other Royal Air Force stores. Later in the year, i.e. October 21st, one company, 1st Battalion, took over the guards on the aerodrome at Mosul. On August 1st the company at Billeh Camp was withdrawn, a Battalion of the Iraq Army took its place.

Beginning of final cutting down. In view of the terms of the Treaty with Iraq, which laid down that the effective strength of the Force was to be 1,250, a gradual reduction down to that number by 1st April, 1933, was put in hand.

This involved the disbandment of a number of the Assyrians, and with a view to helping them a scheme for settlement in the Baradost area was put forward by the High Commissioner and approved by the Iraq Government. To help this scheme, one company of the 2nd Battalion marched to the Baradost on August 28th and camped at Haruna until the end of November. But the prospective settlers did not turn up, for many reasons of their own, and Sheikh Ahmed of Barzan made matters more difficult by having a private war in the area against Sheikh Rashid of Lolan. The company rejoined the Battalion at Diana at the end of November.

1932. One Company to Basrah.

On the 16th February, 1932, one company of the 2nd Assyrian Battalion left Sulaimani for Basrah to take over guard duties at Margil and Shaibah. These duties were taken over on 19th February, 1932.

During the year other changes are due, and by the end of the year the name " Levies " ends, and the new " Air Defence Force " comes into being.

The work of the Levies is done. If it has been done well, it is for others to say.

APPENDIX I

BRITISH AND INDIAN PERSONNEL IN RECEIPT OF AWARDS WHILST SERVING WITH THE IRAQ LEVIES

Name.	Date.	Award.
Major C. A. Boyle	1920	D.S.O., Mentioned in Despatches
R.S.M. W. J. Simmons	1920	Military Medal
R.S.M. E. G. Walker	1920 1926	Mentioned in Despatches M.B.E.
Capt. C. H. Wyncol	1920	Mentioned in Despatches
Capt. H. A. Foweraker	1920	Mentioned in Despatches
Lieut. T. F. Davies	1920	Mentioned in Despatches
Capt. H. P. F. Young	1921	O.B.E., Mentioned in Despatches
Capt. C. E. Simpson	1921	M.B.E.
Capt. J. O. Watt	1921 1926	M.B.E. O.B.E.
Capt. C. E. Littledale	1921	M.C., Mentioned in Despatches
Capt. O. M. Fry	1925	M.C.
Lieut. P. J. T. Baddiley	1925	M.B.E.
Col. Comdt. H. T. Dobbin, D.S.O.	1925	C.B.E.
Col. C. R. Barke, T.D.	1925 1928 1929	C.B.E. Mentioned in Despatches Mentioned in Despatches
Major F. W. Hall	1925	M.B.E.
Major J. M. L. Renton	1925 1927	M.B.E. O.B.E.
Lieut. R. Hart	1926	M.B.E.
C.S.M. H. J. Edwards	1926	M.S.M.
Lt.-Col. G. C. M. Sorel-Cameron	1926 1929	C.B.E. Mentioned in Despatches
Capt. R. Merry	1926	O.B.E.
Major J. W. Malcolm, M.C.	1926	O.B.E.
Major R. J. R. Freeth, R.A.	1926	O.B.E.
Lieut. A. G. Fairrie	1926	M.B.E.
Lieut. H. M. Curteis	1927	M.C.
Lt.-Col. L. W. Alexander	1927	C.B.E.
Major V. R. Guise, M.C., R.A.	1927	O.B.E.
Lieut. E. V. Packer	1927	M.B.E.
R.S.M. J. G. Venning	1927	M.B.E.
R.S.M. A. MacGregor	1927	M.B.E.
C.S.M. C. A. G. Wilde	1927	M.B.E.
Jemadar Sayed Gul Akbar Shah	1927	M.B.E.
Vet. Asst. Surgeon Nur Mohammed	1927	M.B.E.
Col. Comdt. J. G. Browne, C.M.G., D.S.O.	1928	C.B.E.
Major E. T. Horner, M.C.	1928	Mentioned in Despatches
Bt. Major R. H. L. Fink, M.C.	1928	Mentioned in Despatches

Name.	Date.	Award.
Lieut. E. G. Buckley	1928	Mentioned in Despatches
Lt.-Col. J. M. Gillatt, D.S.O.	1928	O.B.E.
Sgt. E. H. Riches	1928	Mentioned in Despatches
Lieut. R. H. Muirhead, R.E.	1929	M.B.E.
Asst. Surgeon A. Backman	1929	M.B.E.
R.S.M. J. Lamb	1930	M.B.E.
Asst. Surgeon R. C. Gale	1930	M.B.E.
C.S.M. H. G. Norton	1930	M.B.E.
Capt. W. E. Parnell	1931	O.B.E.
C.S.M. J. Higgins	1931	M.B.E.

APPENDIX II

NATIVE PERSONNEL IN RECEIPT OF AWARDS WHILST SERVING WITH THE IRAQ LEVIES

Name.	Date.	Award.
Yuzbashi Hasoon Ibn Falayfil	1920	M.M.
Mulazim awal Shawkat Ibn Hussain	1920	M.M.
Mulazim Thani Yunis Ibn Hussain	1920	M.M.
No. 50 Chaoush Obaid Ibn Radha	1920	M.M.
No. 21 Bash Chaoush Sha " ala " Mohann	1920	M.M.
No. 6 Bash Chaoush Hussain Ibn Sassim	1921	Medal of O.B.E.
No. 12 Sowar Attaya Ibn Rekayis	1921	Medal of O.B.E.
No. 46 Sowar Hussain Ibn Salman	1921	Medal of O.B.E.
No. 85 Ombashi Saaid Ibn Amaru	1921	Medal of O.B.E.
Yuzbashi Yunis	1921	Medal of O.B.E.
No. 180 Chaoush Abbas Ibn Shawaish	1921	Medal of O.B.E.
No. 99 Ombashi Saleh Ibn Khalaf	1921	Medal of O.B.E.
No. 550 Ombashi Niga Ibn Hamayid	1921	Medal of O.B.E.
Asst. Levy Officer Muzaffar Beg Ibn Saleh	1921	Medal of O.B.E.
Mulazim Thani Haider Ibn Hamid	1921	Medal of O.B.E.
No. 133 Ombashi Khomerad	1921	Medal of O.B.E.
Rab Emma Daniel Ismail	1922	Medal of O.B.E.
Rab Emma Shain Gewergis	1926	Medal of O.B.E.
Rab Khamshi Zia Gewergis	1926	Medal of O.B.E.
Rab Emma Ozario Tamras	1926	M.C.
Rab Khamshi Shlimon Sliwo	1926	M.C.
Zabit III Abid Abdullah	1926	M.C.
Zabit III Majid Mousa	1926	M.C.
No. 55494 L./Cpl. Misho Miro	1927	M.M.
Rab Khamshi Eshu Saper	1928	M.C.
No. 65347 C.Q.M.S. Baitu Markus	1928	M.M.
No. 55416 Cpl. Barkhu Bobo	1928	Mentioned in Despatches
Rab Khaila David D'Mar Shimun	1928	Honorary M.B.E.
Rab Emma Malam Oda	1929	Mentioned in Despatches
Rab Khamshi Barkhu Hormis	1929	Mentioned in Despatches
No. 55994 Pte. Khaninia Yakub	1929	Mentioned in Despatches

INDEX

C

D

U

V

W

Y

Z

www.ingramcontent.com/pod-product-compliance
Ingram Content Group UK Ltd.
Pitfield, Milton Keynes, MK11 3LW, UK
UKHW051128260726
13967UKWH00010B/2928